Hemoglobin Function in Vertebrates
Molecular Adaptation in Extreme and Temperate Environments

Springer

Milano
Berlin
Heidelberg
New York
Barcelona
Hong Kong
London
Paris
Singapore
Tokyo

G. di Prisco • B. Giardina • R.E. Weber

Hemoglobin Function in Vertebrates

Molecular Adaptation in Extreme and Temperate Environments

Springer

G. di Prisco
Institute of Protein Biochemistry
and Enzymology
CNR Naples, Italy

B. Giardina
Institute of Chemistry
Faculty of Medicine
Catholic University 'Sacro Cuore', Rome, Italy

R.E. Weber
Center for Respiratory Adaptation (CRA)
Department of Zoophysiology
Institute of Biological Sciences
University of Aarhus, Aarhus, Denmark

Cover figure: Ribbon representation of the molecular structure of skua (*Catharacta maccormicki*) Hb 1. The $\alpha_1\beta_1$ and $\alpha_2\beta_2$ dimers are shown in light and dark red, respectively; hemes are in blu; IHP is in grey.

Springer-Verlag is a company in the BertelsmannSpringer publishing group.

© Springer-Verlag Italia, Milano 2000

ISBN 88-470-0107-2

Cover design: Simona Colombo, Milan, Italy
Typesetting: Graphostudio, Milan, Italy
Printing and binding: Staroffset, Cernusco s/N, Milan, Italy

SPIN: 10762361

QP
96
.5
H4485
2000

Preface

The annual Congress of the Italian Biochemical and Molecular Biology Society (SIB) was held in September 1999 in Alghero, Sardegna, Italy. The programme envisaged a symposium on molecular adaptations of haemoglobin function in vertebrates. Haemoglobin specialists from several countries were invited to speak at the symposium and paved the way for wide-ranging and stimulating discussions.

The symposium contributions have been collected together in this volume. The structure/function relationship in haemoglobins from vertebrates (fishes populating temperate and polar environments, diving birds, marine and terrestrial mammals) has been tackled from many angles, focusing on the adaptation of the oxygen-transport system to the constraints dictated by the environment. Eleven articles review some of the most recent developments of the studies on this ancient oxygen-transport protein, characterized by high conservation during evolution.

The volume offers the reader an updated, state-of-the-art summary of a field that is enjoying a true renaissance. Covering the topic from several viewpoints, the volume includes protein chemistry (amino acid sequence, secondary, tertiary and quaternary structures, thermodynamics of oxygen-binding features), molecular biology (globin gene structure, sequence, organization, expression and regulation) and evolution. In this representation of effective multidisciplinary and multinational collaborative efforts, reference is available to a wide range of disciplines and biological systems. The tools of the investigators comprise advanced and powerful methodologies developed in recent years, e.g. site-directed mutagenesis for exchanging amino acid residues, assessing their specific roles (and discovering new ones) and involvement in conformational and allosteric transitions; X-ray structural analyses and molecular modelling to elucidate molecular stereochemistry and modifications in structure/function relationships responsible for the adaptive strategies developed in response to environmental constraints. Combined with the well-defined physiological roles of haemoglobins, a wealth of opportunities are given for identifying links between molecular structures and the significance and expression at the organismic level. Insight into these mechanisms is important in its own right and is also a prerequisite for a host of applied fields (extending from medical practice and animal husbandry to molecular modelling and drug design).

It is not by chance that Sardinia was chosen as the venue of the meeting. In fact, the island currently enjoys the status of being the origin and location of the scientific activity of a number of local research teams, which ensures a continuous flow of exciting findings on haemoglobins of marine and terrestrial vertebrates. The volume accurately reflects the degree of interaction and the number of productive collaborations which have been steadily developing among the authors during recent years.

We would like to express our gratitude to the speakers, authors and referees of the papers collected together in this volume, and also (last but not least) to Prof Bruno Masala and Prof Laura Manca who organized this stimulating symposium in beautiful surroundings. The financial support of the University of Sassari, of the Fondazione Banco di Sardegna, and of PNRA (Italian National Programme for Antarctic Research) is gratefully ackowledged.

May 2000

G. di Prisco
B. Giardina
R. E. Weber

Contents

The Hemoglobin Polymorphism in Sardinian Goats: Nucleotide
Sequence and Frequency of β^A, β^D, $\beta^{\text{D-Malta}}$, and β^E Globin Genes 97
M. PIRASTRU, M. PALICI DI SUNI, G.M. VACCA, P. FRANCESCHI, B. MASALA, L. MANCA

The Organization of the β-Globin Gene Cluster and the Nucleotide Sequence
of the β-Globin Gene of Cyprus Mouflon (*Ovis gmelini ophion*) 109
E. SERRERI, E. HADJISTERKOTIS, S. NAITANA, A. RANDO,
P. FERRANTI, M. CORDA, L. MANCA, B. MASALA

List of Contributors

CLEMENTI M.E., 77, 91
CORDA M., 61, 71, 109
DE ROSA M.C., 71
DEIANA A.M., 61
DETRICH H.W., III, 39
DI PRISCO G., 1, 51, 71, 83
FAIS A., 61, 71
FERRANTI P., 109
FICARRA S., 77, 91
FRANCESCHI P., 97
GALTIERI A., 77, 91
GIARDINA B., 1, 71, 77, 83, 91
HADJISTERKOTIS E., 109
LANIA A., 77
LUPI A., 77, 91
MANCA L., 97, 109
MASALA B., 97, 109

NAITANA S., 109
OLIANAS A., 61, 71
PALICI DI SUNI M., 97
PELLEGRINI M., 61, 71
PIRASTRU M., 97
PISANO A., 61
RANDO A., 109
RICCIO A., 83
RUSSO A.M., 77
SALVADORI S., 61
SANNA M.T., 61
SERRERI E., 109
TAMBURRINI M., 51, 71, 83
TELLONE E., 77
VACCA G.M., 97
WEBER R.E., 23

Molecular Aspects of Temperature Adaptation

G. di Prisco[1], B. Giardina[2]

Introduction

Organisms living in extreme environments, such as Arctic and Antarctic polar regions, are exposed to strong constraints. One of these is temperature, often a stringent driving factor. Consequently, their evolution has included special adaptations, some of which have caused special modifications that have significant implications at the biochemical, physiological and molecular levels. Along these lines, scrutiny of macromolecules, e.g. proteins, provides an opportunity to establish structure-function relationships which may be involved in a specific adaptation to different physiological requirements.

In this chapter the molecular aspects of some temperature adaptations characterising polar marine and terrestrial organisms are summarised, paying special attention to oxygen-transport haemoproteins.

Oxygen Transport

Oxygen carriers are one of the most interesting systems for studying the interrelationships between environmental conditions and molecular evolution. Haemoglobin (Hb), a direct link between the exterior and the requirements of the body, experienced major evolutionary pressure to adapt its functional features. In order to ensure an adequate supply of oxygen to the entire organism, Hbs have developed a common molecular mechanism based on ligand-linked conformational change in a multi-subunit structure. They generally exhibit a marked degree of co-operativity between oxygen-binding sites (homotropic interactions), which enables maximum oxygen unloading at relatively high oxygen tension. In the simplest model, co-operativity in oxygen binding is

[1]Institute of Protein Biochemistry and Enzymology, CNR, Naples, Italy
[2]Institute of Chemistry, Faculty of Medicine, Catholic University 'Sacro Cuore', Rome, Italy

G. di Prisco, B. Giardina, R.E. Weber (Eds)
Hemoglobin Function in Vertebrates.
Molecular Adaptation in Extreme and Temperate Environments
© Springer-Verlag Italia 2000

achieved through conformational transition between a low-affinity T state and a high-affinity R state, which accounts for the sigmoidal shape of the ligand-binding curve. These two extreme conformational states are also involved in the modulation of the oxygen affinity brought about by several effectors (heterotropic interactions). Under physiological conditions a given effector may preferentially bind to the low-affinity or to the high-affinity state of the protein, thereby lowering or enhancing the overall oxygen affinity of the molecule.

Within the framework of this common mechanism the respiratory proteins of polar organisms have acquired adaptive mechanisms to meet special needs. One of our research areas is the structure/function of haemoproteins and the search for a correlation between haematology and the extreme conditions of the polar environments [1-11]. This has prompted an investigation of the relationship between the molecular structure and the oxygen-binding properties, on the one hand, and the ecological constraints, on the other. In view of the role of temperature in modifying the oxygenation-deoxygenation cycle in respiring tissues, thermodynamic analysis of the connection between the binding of heterotropic effectors and the reaction with oxygen deserves special attention.

The Antarctic

Paleogeographic events are the necessary background for understanding the evolutionary history and the adaptations of Antarctic fish. In the late Precambrian, 590 million years ago (Ma) and for 400 million years, during the Paleozoic and part of the Mesozoic through the Jurassic, Antarctica was the central part of the supercontinent Gondwana. Fragmentation during the Cretaceous and continental drift took Antarctica to the present position 65 Ma, at the beginning of the Cenozoic. About 25 Ma, in the Oligocene-Miocene transition, the opening of the Drake passage produced the development of the Antarctic polar front, where the surface layers of the north-flowing Antarctic waters sink beneath the less cold, less dense sub-Antarctic waters. Although sea ice may have already been present at the end of the Eocene (40 Ma), extensive ice sheets, with a periodic occurrence every 1-3 Ma, did not form until after the middle Miocene (14 Ma), and the latest expansion of the ice sheet began 2.5 Ma, in the Pliocene. With the reduction of heat exchange from the northern latitudes, the environment proceeded to cool to the present extreme conditions. The Antarctic ocean became gradually colder and seasonally ice-covered.

Physiological and Biochemical Adaptations in Antarctic Fish

The modern Antarctic fish fauna is largely endemic and, unlike the populations of the other continental shelves, is dominated by the single suborder Notothenioidei. Ninety-five of the 174 species living on the shelf or upper slope of the Antarctic continent are notothenioids [12, 13]. Indirect indications suggest that notothenioids appeared in the early Tertiary, filling the ecological void on the shelf left by most of the other fish fauna (which became locally extinct during maximal glaciation), and began to diversify in the middle Tertiary. Less competition and increasing isolation favoured speciation. Notothenioids fill a varied range of ecological niches normally occupied by taxonomically diverse fish communities in temperate waters. The suborder comprises eight families (Table 1). They are red-blooded, with the remarkable exception of Channichthyidae (the most phyletically derived family), the only known vertebrates whose pale whitish blood is devoid of Hb [14].

Table 1. The families of the suborder Notothenioidei [12, 28]

Family	Antarctic species	Non-Antarctic species	Total
Bovichtidae	1	9	10
Pseudaphritidae	0	1	1
Nototheniidae	34	14	48
Eleginopidae	0	1	1
Harpagiferidae	6	0	6
Artedidraconidae	24	0	24
Bathydraconidae	15	0	15
Channichthyidae	15	0	15
Total	95	25	120

Although in this environment fish from temperate waters would rapidly freeze, the oxygen-rich Antarctic waters do support a wealth of marine life.

The ensemble of physiological/biochemical adaptations (e.g. freezing avoidance; efficient enzymatic catalysis and cytoskeletal polymer assembly; decreased blood viscosity) was developed in the last 20-30 million years, during increasing isolation in the cooling seas. Some of these adaptations are a unique characteristic of polar fish, and fish now live in isolation south of the Polar Front, a natural barrier to migration in both directions and a key factor for evolution. These fish had to cope with temperatures below the freezing point of the body fluids and with high oxygen concentrations, adjusting specifically to the environment such that they are intolerant of warmer temperatures. In addition to the low and constant temperature, seasonality of other variables such as light and productivity may influence biological processes such as feeding, growth and reproduction [15].

Blood and the Oxygen-Transport System

The haematological features of many Antarctic Notothenioidei have been extensively investigated in the past few decades. In fact, notothenioids are by far the most thoroughly characterised group of fish in the world. Correlations with fish activity patterns have been proposed, although the dramatic effect that stress invariably exerts on the haematological parameters calls for extreme caution [16, 17]. Our studies on the oxygen-transport system have addressed 38 out of a total of 80 red-blooded Antarctic species so far [4] and are aimed at correlating sequence, multiplicity and oxygen binding with ecological constraints and at obtaining phylogenetic information on evolution. This highly representative number encompasses all major families and also, for comparative purposes, two species of non-Antarctic notothenioids.

Antarctic fish clearly differ from temperate and tropical species in having a reduced erythrocyte number and Hb concentration in the blood. The subzero seawater temperatures would greatly increase the viscosity of blood, with potentially negative physiological effects; however, this increase is offset by reducing or eliminating erythrocytes and Hb. This adaptation has reduced the amount of energy needed for circulation.

Hb-less Channichthyidae represent the extreme of this trend. In these fish, Hb has not been replaced by another carrier and the oxygen-carrying capacity of blood is only 10% of that of red-blooded fish. However, these fish are not at all at a disadvantage from the lack of Hb. The physiological adaptations enabling channichthyids to prosper without Hb include: low metabolic rate; large, well-perfused gills; large blood volume, heart, stroke volume and capillary diameter; and cutaneous respiration. In addition to reducing the metabolic demand for oxygen, low temperatures increase its solubility in the plasma, so that more oxygen can be carried in physical solution. The co-existence of Hb-less and naturally cytopenic red-blooded species suggests that the need for an oxygen carrier in a stable, cold environment is also reduced in red-blooded fish. In fact, functional incapacitation of Hb (by means of exposure to carbon monoxide) and reduction of the haematocrit levels from 8%-15% to 1% in cannulated specimens of *Trematomus bernacchii* [17, 18] caused no discernible harm in the absence of metabolic challenges. Similar to channichthyids, red-blooded fish can carry routinely needed oxygen dissolved in the plasma.

The evolutionary loss of the respiratory pigments is a highly specialised condition which raises several questions. If the Hb-less state is adaptive for one family, why not also for the others living in the same habitat? No unequivocal answer has yet been found, and Wells [19] has suggested that the Hb-less state may be non-adaptive. Another question is: what happened to the α- and β-globin genes? Relieved of selective pressure for expression, they may have diverged from those of red-blooded notothenioids, or been lost altogether. The characterisation of globin DNA sequences from several species of red-blooded and Hb-less notothenioids has indicated that three channichthyids, spanning the clade from primitive to advanced genera, share retention of α-globin-related, transcriptionally inactive DNA sequences in their genomes and apparent loss, or rapid mutation, of β-globin genes [20, 21]. This common pattern suggests that loss of globin-gene expression is a primitive character, established in the common ancestral channichthyid approx. 25 Ma prior to diversification within the clade. Deletion of the β-globin locus of the ancestor may have been the primary event leading to the Hb-less phenotype; the α-globin gene(s), no longer under selective pressure for expression, would then have accumulated mutations leading to loss of function without, as yet, complete loss of sequence information.

A common feature of endemic Antarctic fish is the markedly reduced Hb multiplicity, and in fact, in the first stages of our research, this fauna seemed rather uniform and simplified. Among notothenioids (Table 2), 34 species [3] have a single major Hb (Hb 1) and often a second, functionally similar minor component (Hb 2, about 5% of the total, usually having the β chain in common with Hb 1). Another component (Hb C) is present at less than 1% in all species. All these fish are sluggish bottom dwellers. This was a first important correlation.

A general trend was found in the amino acid sequences. Those of major and minor Hbs cluster in two groups; in each group, identity is high (73%-99% and 84%-100%, respectively). However, the identity between major and minor Hbs is lower, ranging between 61% and 73%. The analysis of sequences, together with the similar functional features of major and minor Hbs in a given species, led us to conclude that minor Hbs are vestigial (perhaps larval) remnants, devoid of physiological significance [1,4].

The effect of pH and endogenous organophosphate effectors on oxygen equilibria and saturation, i.e. the Bohr [22] and Root [23] effects, have been investigated in 30 of the 34 notothenioids shown in Table 2, including two (a pseudaphritid [24] and a nototheniid) non-Antarctic species (Table 3), plus an additional Antarctic one. The haematocrit, erythrocyte number, Hb concentration and MCHC of non-cold-adapted *Notothenia angustata* are higher than those of Antarctic nototheniids, as expected in a fish of lower latitudes; but Hb multiplicity and the structural/functional features closely resemble those of the Antarctic species of the same family and suborder [25]. The most striking similarity concerns the primary structures, which compare in identity with those of Antarctic *Notothenia coriiceps* Hbs at a level of 93% in the β chains and of 99% in the α chains of Hb 1; the α chains of Hb 2 are identical. This level of identity between a cold-adapted and a non-cold-adapted species of the same family is the highest ever found among notothenioid Hbs. This evidence and the finding of antifreeze genes in the genome of *N. angustata* (Cheng, personal communication) suggest that this fish was cold adapted prior to its quite recent migration from Antarctic to temperate waters.

As analysed in detail by Tamburrini et al. (this volume), three nototheniids (active and cryopelagic *Trematomus newnesi* and *Pagothenia borchgrevinki*, and *Pleuragramma antarcticum*, a pelagic, sluggish but migratory fish) have a widely different life style from the 35 sluggish benthic species listed in Tables 2 and 3. They do not follow the pattern of low multiplicity, but have three to five

Table 2. Haemoglobin (Hb) in the blood of Antarctic and non-Antarctic (*Notothenia angustata* and *Pseudaphritis urvillii*) Notothenioidei

Family	Species	Hb components
Pseudaphritidae		
	P. urvillii	Hb 1 (95%), Hb 2 (5%)
Nototheniidae		
	N. coriiceps	Hb 1 (95%), Hb 2 (5%)
	N. rossii	Hb 1 (95%), Hb 2 (5%)
	N. angustata	Hb 1 (95%), Hb 2 (5%)
	N. nudifrons	Hb 1 (95%), Hb 2 (5%)
	N. larseni	Hb 1 (95%), Hb 2 (5%)
	G. gibberifrons	Hb 1 (90%), Hb 2 (10%)
	T. hansoni	Hb 1 (95%), Hb 2 (5%)
	T. bernacchii	Hb 1 (98%) (Hb 2?)
	D. mawsoni	Hb 1 (98%) (Hb 2?)
	A. mitopteryx	one Hb (99%)
	T. nicolai	Hb 1 (95%), Hb 2 (5%)
	T. pennellii	Hb 1 (95%), Hb 2 (5%)
	T. loennbergi	Hb 1 (95%), Hb 2 (5%)
	T. eulepidotus	Hb 1 (95%), Hb 2 (5%)
	T. lepidorhinus	Hb 1 (95%), Hb 2 (5%)
	T. scotti	Hb 1 (95%), Hb 2 (5%)
Bathydraconidae		
	C. mawsoni	Hb 1 (95%), Hb 2 (5%)
	R. glacialis	Hb 1 (90%), Hb 2 (10%)
	P. charcoti	one Hb (99%)
	G. acuticeps	one Hb (99%)
	B. marri	one Hb (99%)
	B. macrolepis	one Hb (99%)
	A. nudiceps	one Hb (99%)
	G. australis	one Hb (99%)
Artedidraconidae		
	A. skottsbergi	one Hb (99%)
	A. orianae	one Hb (99%)
	A. shackletoni	one Hb (99%)
	H. velifer	one Hb (99%)
	P. scotti	one Hb (99%)
	Pogonophryne sp 1	one Hb (99%)
	Pogonophryne sp 2	one Hb (99%)
	Pogonophryne sp 3	one Hb (99%)
Harpagiferidae		
	H. antarcticus	one Hb (99%)

The blood of all species contains traces (less than 1%) of Hb C.

Table 3. Regulation by pH and physiological effectors of oxygen binding of Hbs of Antarctic and non-Antarctic (*N. angustata* and *P. urvillii*) Notothenioidei

Family	Species	Bohr and Root effects; effect of organophosphates
Pseudaphritidae		
	P. urvillii	Strong in Hb 1, Hb 2
Nototheniidae		
	N. coriiceps	Strong in Hb 1, Hb 2
	N. rossii	Strong in Hb 1, Hb 2
	N. angustata	Strong in Hb 1, Hb 2
	G. gibberifrons	Strong in Hb 1, Hb 2
	T. hansoni	Strong in Hb 1, Hb 2
	T. bernacchii	Strong in Hb 1
	D. mawsoni	Strong in Hb 1
	A. mitopteryx	Root, absent; Bohr, Weak
	T. nicolai	Strong in Hb 1, Hb 2
	T. pennellii	Strong in Hb 1, Hb 2
	T. loennbergi	Strong in Hb 1, Hb 2
	T. eulepidotus	Strong in Hb 1, Hb 2
	T. lepidorhinus	Strong in Hb 1, Hb 2
	T. scotti[a]	Strong in Hb 1, Hb 2
Bathydraconidae		
	C. mawsoni	Strong in Hb 1, Hb 2
	R. glacialis	Strong (in haemolysate)
	P. charcoti	Strong
	G. acuticeps	Absent
	B. marri	Strong
	B. macrolepis[a]	Strong, only with ATP
	A. nudiceps[a]	Strong
	G. australis[a]	Strong
Artedidraconidae		
	A. orianae	Weak (Root only with ATP)
	A. shackletoni[a]	Weak, only with ATP
	H. velifer	Weak
	P. scotti	Weak (Root only with ATP)
	D. longedorsalis[a]	Weak (in haemolysate)
	Pogonophryne sp 1	Weak (Root only with ATP)
	Pogonophryne sp 2	Weak (Root only with ATP)
	Pogonophryne sp 3	Weak (Root only with ATP)

[a]The Bohr effect was not measured.

functionally distinct Hb components (Table 4). Each species has a unique oxygen-transport system, and each system appears adjusted to the fish specific mode of life.

Table 4. Antarctic Notothenioidei (family Nototheniidae) with higher Hb multiplicity and with functionally distinct components

Species	Hb components	Bohr effect	Root effect	Effect of organophosphates
T. newnesi (two major Hbs)				
	Hb C (20%)	Strong	Strong	Strong
	Hb 1 (75%)	Weak	Absent	Absent
	Hb 2 (5%)	Weak	Absent	Absent
P. antarcticum [a] (3 major Hbs)				
	Hb C (traces)	Strong	Strong	Strong
	Hb 1 (30%)	Strong	Strong	Strong
	Hb 2 (20%)	Strong	Strong	Strong
	Hb 3 (50%)	Strong	Strong	Strong
P. borchgrevinki (one major Hb)				
	Hb C (traces)	Strong	Strong	Strong
	Hb 0 (10%)	Strong	Strong	Strong
	Hb 1 (70%)	Weak	Weak	Absent
	Hb 2 (10%)	Weak	Weak	Weak
	Hb 3 (10%)	Weak	Weak	Weak

[a]The Hbs of *P. antarcticum* differ thermodynamically (they have different values of heat of oxygenation).

T. newnesi is the only species in which Hb C is not present in traces and has two major, functionally distinct Hbs [26]. Such a Hb system may be required by this more active fish to ensure oxygen binding at the gills and controlled delivery to tissues, even though the active behaviour produces acidosis. With three major Hbs, *P. antarcticum* has the highest multiplicity among notothenioids. These Hbs display almost identical effector-enhanced Bohr and Root effects, but differ thermodynamically in the values of oxygenation heat [27]. The low enthalpy change is interpreted as adaptive, since a lower heat input is needed to drive oxygenation. Temperature-regulated oxygen affinity may reflect highly refined molecular adaptation to pelagic life and to migration across water masses with significant temperature differences and fluctuations. *P. borchgrevinki* has five Hbs [25] with different pH and organophosphate regulation, as well as different oxygenation heats, perhaps indicative of the most specialised oxygen-transport system among the notothenioids.

Seven species of the family Artedidraconidae were found to have a single Hb, with weak Bohr and Root effects. Two Hbs were thoroughly investigated and found to lack oxygen-binding co-operativity, in a way similar to that of the ancestral haemoproteins of primitive organisms [28], raising intriguing questions on the mode of function of multisubunit molecules and on evolutionary implications.

Some Other Adaptive Specialisations

Neutral Buoyancy. All Antarctic fish lack the swim bladder, which would produce neutral buoyancy and save energy during locomotion and displacement in the water column since a neutrally buoyant fish is weightless. In notothenioids (a primarily benthic group, although some species have become seasonally or even permanently adapted to pelagic life) evolution compensated for the lack of the swim bladder by inducing modifications in a variety of body systems, facilitating rapid vertical migration with maximal energy conservation. This was attained through a combination of efficient strategies [13], i.e. reduction of bone and scale mineralisation (actually, many species are scaleless); substitution of bone with cartilage (less dense); storage of lipid (providing static lift); and production of a subdermal layer of watery, gelatinous tissue.

Freezing Avoidance. The average water temperature in the coastal part of the Antarctic ocean is -1.87°C, the equilibrium temperature of ice and sea water, well below the freezing temperature of a typical marine teleost hyposmotic to seawater. A few species avoid freezing by supercooling. However, the supercooled state is metastable and requires the absence of contact with ice; although in deeper water ice formation is impaired by the hydrostatic pressure effect, currents may carry ice crystals, making supercooling a dangerous strategy since even partial freezing causes death. In Notothenioidei, freezing is avoided by lowering the freezing point of blood and other tissue fluids. In addition to NaCl, the freezing point depression is provided by solutes in the colloidal fraction of the fluid. In most Antarctic fish species these are glycopeptides (AFGPs) containing a repeating unit of three residues in the sequence [Ala-Ala-Thr]n; a disaccharide is linked with each Thr. In the smaller peptides, Pro periodically substitutes Ala at position 1 of the tripeptide. AFGPs are synthesised in the liver, secreted into the circulatory system and distributed into the extracellular fluids [29, 30]. In a most elegant development of this research, it was shown that the AFGP gene of notothenioids, but not of Arctic cod, evolved from a pancreatic trypsinogen gene. Nonetheless, the two unrelated types of fish essentially have the same AFGPs: a rare case of convergent evolution in the protein sequence [31].

Cytoskeletal Polymers: Tubulins and Actins. Tubulins are proteins which, together with microtubule-associated proteins, assemble and form subcellular structures (microtubules). These are a major component of the cytoskeleton of eukaryotic cells and participate in many processes, e.g. mitosis, nerve growth and intracellular transport of organelles. The in vitro assembly of microtubules is temperature-sensitive; in temperate fish, mammals and birds, tubulins associate at about 37°C, but disassemble at temperatures as low as 4°C. In notothenioids, however, adaptive changes in the tubulin molecular structure have made these proteins able to stay polymerised at up to -2°C. Polymerisation relies on entropy-generating interactions, and, more specifically, on an increase in hydrophobic interactions between the molecular domains involved [32].

Skeletal muscle actins are another interesting example of protein polymerisation [33]. Self-assembly is temperature-dependent. It has been suggested that the stabilisation of actin filaments in cold-adapted fish depends on exothermic polar bonds rather than on endothermic hydrophobic interactions.

Thus, cytoskeletal polymer assembly at low temperatures is regulated by at least two adaptive strategies: (1) modification of the bond types at subunit contacts gives an entropy of association sufficient to overcome unfavourable enthalpy changes; (2) a preponderant role of bonds making negative contributions to the enthalpy of polymerisation limits the destabilising enthalpy changes.

Enzyme Catalysis. The effect of temperature on enzyme stability and activity calls for thermodynamic analysis. Some Antarctic fish enzymes are more labile than the corresponding mammalian ones [16]. In others, the differences in heat denaturation are either non-existent or very small [34, 35] and suggest conservation of the molecular structure during evolution. At least two types of adaptation can account for high catalytic rates at low temperatures [36], (1) a higher intracellular enzyme concentration (an increased number of catalytic sites compensates for the temperature-induced lower rate per site), and (2) a higher inherent catalytic activity per active site (higher activity is achieved by means of fewer molecules of a more efficient enzyme). The two types often co-exist. For example, thermodynamic analysis suggests that glucose-6-phosphate dehydrogenase (G6PD) from the blood of two nototheniods is a "better enzyme" than G6PD from temperate fish [34]. However, the highly increased amount of G6PD in the few cells of the Hb-less blood of the channichthyid *Chionodraco hamatus* strongly suggests temperature compensation via the synthesis of a larger number of G6PD molecules as well.

The Arctic

Specialisations in the blood and in the oxygen-transport system were also developed by other organisms which experience low temperatures, such as Arctic mammals (reindeer, musk ox, whale, brown bear [5-7, 37]), crustaceans (krill [8]), cephalopods (squid [9]). The selected cases include organisms which, although not living at the extreme latitudes of the northern hemisphere (e.g. penguin [10] and turtle [11]), illustrate how evolution can alter the overall thermodynamics to optimise the oxygenation-deoxygenation cycle in dependence of the physiological needs of the particular organism. The cases of krill and squid haemocyanins widen the scope of the scheme, indicating that the molecular mechanisms of temperature adaptations are not substantially different in the two classes of proteins.

Mammals: Ruminants, Bear and Whale

Hb from Arctic and sub-Arctic ruminants (reindeer, musk ox and cervus), under physiological conditions, is characterised by an overall oxygenation enthalpy (ΔH) that is much less exothermic than that of human Hb and other mammalian Hb (Table 5). The physiological implication of this result may become apparent considering the very low habitat temperature (down to -40°C) experienced by these animals during the year. Thus, as deoxygenation is an endothermic process, in the peripheral tissues, where the temperature may be as much as 10°C lower than in the lungs and the deep core of the organism, oxygen delivery would be drastically impaired if the molecule were not characterised by a small ΔH, namely by a slight temperature dependence of the oxygen binding. That the protein molecule possesses peculiar features is clearly shown in reindeer Hb, in which oxygen binding has been investigated in great detail by a set of experiments carried out as a function of temperature [5]. Typical results are shown in Fig. 1. The data, presented as a Hill plot, extend over a saturation range broad enough to permit evaluation of a number of thermodynamic parameters. One feature which shows up very clearly is the strong temperature dependence of the shape of the binding curve. In fact, a temperature increase brings about a great decrease in the association constant for the binding of the first oxygen molecule (indicated by the lower asymptote), without significantly affecting that for the binding of the last oxygen molecule (indicated by the upper asymptote). This is mainly due to the difference in the overall heat of oxygenation relative to the T and the R states of the molecule. Thus, while ΔH of oxygen binding to the T state is strongly exothermic, that of the R state is close to zero. This dramatic difference in the thermodynamics of the two conformational states of reindeer Hb results in a strong dependence of the temperature effect on the degree of oxygen saturation of the pro-

Table 5. Overall heat of oxygenation of some ruminant Hb and, for comparison, of human and horse Hb under the following conditions: 0.1 M Hepes, plus 0.1 M NaCl in the presence of 3 mM DPG at pH 7.4

Species	ΔH (kcal mol^{-1})
Reindeer	-3.29
Musk ox	-3.48
Cervus	-3.08
Horse	-6.78
Man	-7.98

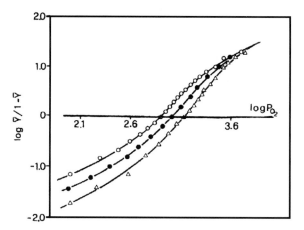

Fig. 1. Effect of temperature on oxygen equilibria of reindeer Hb measured in 4% carbon dioxide, 0.05 M Tris-HCl pH 7.4 at 10°C (*open circles*), 15°C (*filled circles*) and 20°C (*open triangles*). Oxygen pressure (Po_2) is expressed in Pa

tein. As a consequence, within the range of oxygen saturation in which the protein is working in vivo, the overall heat of oxygenation is very low and approaches 0 as the fractional saturation moves toward 1.0. On the whole, the Hb of these Arctic ruminants may be regarded as a beautiful example of molecular adaptation to extreme environmental conditions.

The data obtained on Hbs from musk ox and cervus indicate that the small ΔH of oxygen binding could be considered an intrinsic property of the molecule, which is the result of an evolutionary adaptation at the molecular level. Hence, in the case of musk ox, the ΔH of oxygenation is at its maximum value (in absolute terms), even if small, just within the physiological pH range and approaching 0 when the pH tends to be more acidic and more alkaline. We may, therefore, exclude a significant involvement of the Bohr protons in determining this unusual ΔH of oxygen binding and hypothesise the effect of some other ions whose presence could be important in vivo for the definition of the overall functional properties of the molecule.

Another interesting example of molecular adaptation is provided by Hb from the brown bear *Ursus arctos* [37], in which chloride and 2,3-diphosphoglyceric acid (2,3-DPG) modulate the oxygen affinity in a synergistic way such that their individual effect is enhanced whenever they are both present in saturating amounts (Fig. 2). The thermodynamic analysis of such a feature indicates that in bear Hb there are two classes of chloride-binding sites, one acting synergistically with 2,3-DPG and another which is likely to overlap with the organophosphate interaction cleft and therefore only fully operative in the absence of 2,3-DPG. The physiological importance of this synergistic effect can be fully appreciated with respect to the relevance of the "chloride shift" in the regulation of oxygen unloading and loading at the level of capillaries.

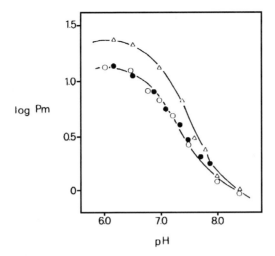

Fig. 2. Effect of pH on the oxygen affinity (in terms of log p_m) of brown bear Hb at 20°C. Buffer conditions: 0.1 M bis-Tris or Tris-HCl plus 0.1 M NaCl in the absence (*open circles*) and presence (*open triangles*) of 3 mM 2,3-DPG. *Filled circles* represent experiments performed in 0.1 M chloride-free Hepes buffer plus different concentrations of 2,3-DPG, which are saturating for HbO_2 at each pH

Moreover, it must be pointed out that ΔH for oxygen binding displays a marked pH-dependence such that, under conditions of acidosis (which is likely to occur during hibernation) the exothermicity of oxygenation is drastically reduced. Therefore, under these conditions, even a decrease in temperature in the peripheral tissues would have essentially no consequences for their oxygen supply. Furthermore, it should be recalled that acidosis is usually associated with hypercloruraemia; hence the peculiar synergistic effect of chloride and 2,3-DPG on the oxygen-binding properties of bear Hb could have physiological relevance in enhancing oxygen release from Hb under hypercloruraemic and acidotic conditions at low temperatures, i.e. during hibernation.

Among mammals living in extreme environments, the whale *Balaenoptera acutorostrata* represents an additional variation on the theme of the relationships between evironmental conditions and functional modulation. Whale Hb [7] is, in fact, characterised by two main features: (1) although the intrinsic temperature sensitivity is extremely high, when the physiological co-factors (chloride, 2,3-DPG and carbon dioxide) are added to the system, the overall heat required for deoxygenation is very similar to that previously observed in the case of Arctic ruminants; (2) the effect of carbon dioxide in decreasing the oxygen affinity of the molecule is very similar to that already known in human Hb. However, this effect progressively diminishes as the temperature increases and appears to be completely abolished at 37°C.

The significance of the low temperature dependence of oxygen binding in whale Hb may become apparent if we consider that, although the major part of the whale's body is covered by a thick isolating layer of blubber, the metabolically very active parts, i.e. the flippers and the huge tail, are not insulated so well. Thus, they are kept at a lower temperature by a countercurrent heat exchanger in order to reduce heat loss. Oxygen unloading in these active regions of the whale body is thus very similar to that occurring in the cold leg muscles of Arctic ruminants. Finally, the lower temperature which blood encounters at the fins and tail, in comparison with the rest of the body, also explains the temperature dependence of the effect of carbon dioxide. In fact, within the core of the organism, at 37°C, carbon dioxide does not display any allosteric effect since its differential binding with respect to oxy- and deoxy-structure is completely abolished. However, carbon dioxide, in turn, facilitates oxygen unloading just at the level of fins and tail, which are regions of great muscular activity and where the temperature is well below 37°C due to the presence of the countercurrent heat exchanger. In conclusion, the combined effects of organophosphates, carbon dioxide and temperature optimise oxygen delivery at all tissues in spite of the marked local heterothermia.

Emperor Penguin

Penguins are Antarctic birds fully committed to aquatic life, being accomplished divers and spending much of their life submerged. For this reason they have developed suitable mechanisms for the maintenance of an adequate oxygen supply to tissues under hypoxic conditions [10]. Emperor penguin (*Aptenodytes forsteri*) Hb displays a Bohr effect which appears well adapted for gas exchange during very prolonged dives. In fact, in the presence of the physiological effectors, Hb is characterised by a Bohr coefficient ($\Delta \log P_{50}/\Delta pH$) which is 50% smaller than that of human Hb in the presence of 2,3-DPG (i.e. -0.35 for penguin vs -0.73 for human Hb). In addition, the strongly reduced Bohr effect shows a substantial shift of the mid point of the transition towards acidic pH values (overall $pK \cong 7.2$ for penguin vs 7.7 for human Hb). In this way, the increase in lactic acid and the concomitant decrease in pH accompanying prolonged dives would not affect the oxygen affinity to any great extent, preserving Hb from an uncontrolled and sudden stripping of oxygen. In other words, during diving, oxygen delivery from penguin Hb would be essentially modulated by the oxygen partial pressure at the level of the specific tissue.

It may be of interest to recall that all the main functional characteristics displayed by penguin Hb (i.e. a small Bohr effect with an overall pK shifted towards acidic pH values) are also shown in Hb from the sea turtle *Caretta caretta*, a reptile well specialised for aquatic life in temperate waters, being able to endure very prolonged dives [11]. The effect of temperature on oxygen binding of both Hbs deserves special attention. In turtle Hb, the overall heat of oxygenation is very low (-1.8 kcal/mol), almost four times lower than the value generally observed in vertebrate Hb and even lower than that observed in

Arctic mammals. Therefore, through this very minor enthalpy change, oxygen delivery becomes essentially independent of the water temperature to which the animal is exposed. In penguin Hb, the peculiar pH dependence of the overall ΔH of oxygen binding (Fig. 3) may have special importance in view of its reproductive behaviour. Thus, the very low ΔH observed at acid pH should be considered in connection to the significant metabolic acidosis which accompanies the 64-day egg incubation performed by the male, which holds the egg on his feet (whose temperature, under these conditions, is close to 0°C) and only lives on stored fat reserves.

Crustaceans and Cephalopods: Krill and Squid

Krill (*Meganyctiphanes norvegica*) and squid (*Todarodes sagittatus*) haemocyanin demonstrates that the molecular mechanisms of temperature adaptations seen in Hb are of general validity, since they also apply to a completely different group of oxygen-transport proteins, i.e. the haemocyanins.

Krill, a crustacean epipelagic organism, lives in North Atlantic waters and is abundant throughout the year in Norwegian fjords. Its haemocyanin is characterised by a relatively low oxygen affinity and high co-operativity of oxygen binding [8]. The oxygen affinity of krill haemocyanin increases markedly as a

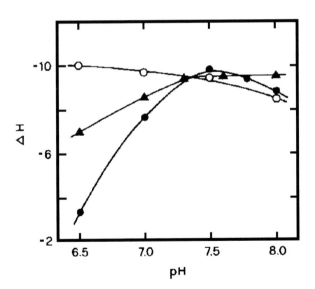

Fig. 3. Overall oxygenation enthalpy (kcal/mol of oxygen) as a function of pH of penguin (*circles*) Hb, in 0.1 M bis-Tris-HCl or Tris-HCl containing 0.1 M NaCl, in the absence (*open symbols*) and presence (*filled symbols*) of 3mM P_6-inositol. The values were calculated from the van't Hoff equation, using data from oxygen equilibrium experiments at different temperatures. Data obtained on pigeon Hb (*triangles*) are reported for purposes of comparison

function of temperature, reflecting a strong endothermic overall heat of oxygenation (ΔH=19 kcal/mol of oxygen at pH 7.6). A possible explanation for the endothermic character of oxygen binding has to be related to an intrinsic property of the macromolecule, or to a specific effect of ions that still needs clarification. In any case, the unusual thermal properties of this haemocyanin remain per se a phenomenon of striking adaptive and evolutionary significance, since they very well fit the behaviour of the animal, which descends during the daytime to depths of 100-400 m and ascends at night to the surface. Hence, the strong endothermic character of oxygen binding should be of great benefit for krill during its feeding excursions when it ascends to the upper layers, where the oxygen availability may be significantly reduced due to lower solubility at higher temperature and to the respiratory activity of plankton.

Another interesting but more complex example is represented by squid haemocyanin [9], whose functional properties have been carefully characterised as a function of temperature and proton concentration. Within the physiological pH range, the concentration of protons mainly affects the high-affinity R state of the molecule without significantly influencing the low-affinity T state. However, the effect of temperature is just the opposite, since the ligand affinity of the T state is greatly affected by temperature changes (oxygen binding is strongly exothermic), whereas that of the R state is almost completely independent of this parameter. This peculiar interplay of pH and temperature effects has a central role in the modulation of oxygen transport, appearing to be of great importance for the physiological requirements of squid. In fact, the differential and opposite effects of protons and temperature minimise the handicap of oxygen unloading induced by a decrease in temperature and the subsequent increase in pH. This is achieved just because temperature and pH affect the lower and the upper asymptotes of the oxygen binding curve, respectively (Fig. 4).

Concluding Remarks

Adaptive evolution of Antarctic fish draws advantage from being staged within a simplified framework (reduced number of variables in a stable environment, dominated by a taxonomically uniform group). Although correlations of molecular data with physiological and biochemical adaptations and with ecology and lifestyle are difficult to identify, the primary importance of this objective is increasingly attracting the interest of scientists. Haematology (a source of basic information on the biochemistry and physiology of a vital function such as respiration) and Hb (a protein whose molecular features have been conserved throughout evolution) have been investigated as potential tools. Studies in a highly representative number of species (38 out of 80 red-blooded Antarctic - and two non-Antarctic - notothenioids) have suggested correlations between lifestyle and Hb multiplicity and functional features. Bottom dwellers have a

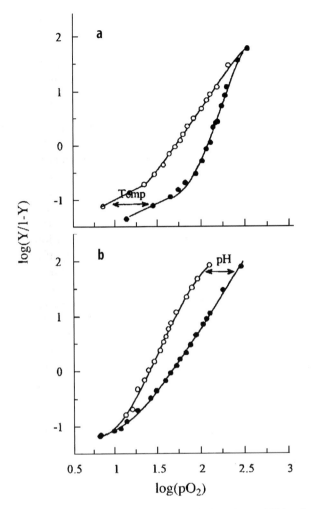

Fig. 4a,b. Oxygen binding curves, expressed by Hill plots, for squid blood as a function of pH and temperature. **a** pH 7.15; 6°C (*left*) and 20°C (*right*) **b** 6°C; pH 7.4 (*left*) and pH 7.15 (*right*)

single major Hb with higher affinity for oxygen; this is hardly unexpected, since the constant physico-chemical conditions of the ocean may have reduced the need for multiple Hbs. The observation that three pelagic and active nototheniids have multiple, functionally distinct Hbs supports this interpretation. In these cases, a link with adaptation to habitat and with lifestyle becomes possible [38].

Perutz [39] has analysed the possible relationships between structure and function of Hb and adaptation. Functional adaptation may have been produced by the gradual accumulation of minor mutations or by substitutions in key positions, and the two alternatives may not be mutually exclusive. The amino acid sequences of Antarctic fish Hb is thus a useful tool in the molecular approach to adaptive evolution. In the construction of phylogenetic trees [40], this study logically implies further extension to non-Antarctic notothenioids of the non-endemic families Bovichtidae, Pseudaphritidae and Nototheniidae, the most ancient of the suborder. In evolutionary history, studies of the globin gene remnants in Hb-less Channichthyidae are also well worth being pursued.

From the thermodynamic point of view, the results outline the role of temperature and of its interplay with some heterotropic ligands in the modulation of respiratory function. In this respect, the energy-saving mechanism displayed by Arctic ruminants is a further, most elegant example of the different strategies adopted during evolution to the transport and release of oxygen at the level of respiring tissues. That the effect of temperature should be studied with respect to the reaction with heterotropic effectors is clearly demonstrated by whale Hb. Here, in fact, temperature controls the modulatory effect of carbon dioxide, switching the differential binding of this ligand on and off depending on the specific region of the organism.

Finally, haemocyanin from krill and squid show that, although respiratory pigments provide good examples of the different chemical strategies adopted during evolution to solve the vital problem of oxygen supply for metabolic demands, the fine modulation of the basic function is achieved, at the molecular level, within the same biophysical framework, namely that of the allosteric control, which in turn may be finely modulated by temperature through specific thermodynamic characteristics of the macromolecule. This concept also seems well represented by bear Hb, which, in order to achieve an adequate oxygen supply, is functionally modulated both through the interaction between different anionic cofactors and temperature effects in a complex but very efficient interplay.

Acknowledgements

This article includes work sponsored by the Italian National Programme for Antarctic Research, carried out thanks to the participation of L. Camardella, V. Carratore, C. Caruso, M.A. Ciardiello, M.E. Clementi, E. Cocca, R. D'Avino, A. Fago, M. Romano, A. Riccio, R. Scatena, M. Tamburrini and the late B. Rutigliano.

References

1. di Prisco G, D'Avino R, Caruso C, Tamburrini M, Camardella L, Rutigliano B, Carratore V, Romano M (1991) The biochemistry of oxygen transport in red-blooded Antarctic fish. In: di Prisco G, Maresca B, Tota B (eds) Biology of Antarctic fish. Springer, Berlin Heidelberg New York, pp 263-281

2. di Prisco G, Tamburrini M, D'Avino R (1998) Oxygen-transport system in extreme environments: multiplicity and structure-function relationship in haemoglobins of Antarctic fish. In: Pörtner HO, Playle RC (eds) Cold ocean physiology. Cambridge University Press, Cambridge, pp 143-165

3. di Prisco G (1997) Physiological and biochemical adaptations in fish to a cold marine environment. In: Battaglia B, Valencia J, Walton DWH (eds) Proc SCAR 6th Biol Symp "Antarctic communities: species, structure and survival". Cambridge University Press, Cambridge, pp 251-260

4. di Prisco G (1998) Molecular adaptations in Antarctic fish hemoglobins. In: di Prisco G, Pisano E, Clarke A (eds) Fishes of Antarctica. A biological overview. Springer, Milano Heidelberg New York, pp 339-353

5. Giardina B, Condo' SG, Petruzzelli R, Bardgard A, Brix O (1990) Thermodynamics of oxygen binding to arctic hemoglobins: the case of reindeer. Biophys Chem 37:281-286

6. Brix O, Bardgard A, Mathisen S, el-Sherbini S, Condo' SG, Giardina B (1989) Arctic life adaptation. II. The function of musk ox hemoglobin. Comp Biochem Physiol 94B:135-138

7. Brix O, Condo' SG, Ekker M, Tavazzi B, Giardina B (1990) Temperature modulation of oxygen transport in a diving mammal (Balaenoptera acutorostrata). Biochem J 271:509-513

8. Brix O, Bougund S, Barunung T, Colosimo A, Giardina B (1989) Endothermic oxygenation of hemocyanin in the krill. FEBS Lett 247:177-180

9. Giardina B, Condo' SG, Brix O (1991) Modulation of oxygen binding in squid blood. In: Vinogradov SN, Kapp OH (eds) Structure and function of invertebrate oxygen carriers. Springer, Berlin Heidelberg New York, pp 333-339

10. Tamburrini M, Condo' SG, di Prisco G, Giardina B (1994) Adaptation to extreme environments: structure-function relationships in Emperor penguin hemoglobin. J Mol Biol 237:615-621

11. Giardina B, Galtieri A, Lania A, Ascenzi P, Desideri A, Cerroni L, Condo' SG (1992) Reduced sensitivity of oxygen transport to allosteric effectors and temperature in loggerhead sea turtle hemoglobin: functional and spectroscopic studies. Biochim Biophys Acta 1159:129-133

12. Gon O, Heemstra PC (eds) (1990) Fishes of the Southern Ocean. JLB Smith Institute of Ichthyology, Grahamstown, South Africa

13. Eastman JT (1993) Antarctic fish biology. Evolution in a unique environment. Academic Press, San Diego

14. Ruud JT (1954) Vertebrates without erythrocytes and blood pigment. Nature 173:848-850

15. Clarke A, North AW (1991) Is the growth of polar fish limited by temperature? In: di Prisco G, Maresca B, Tota B (eds) Biology of Antarctic fish. Springer, Berlin Heidelberg New York, pp 54-69

16. Macdonald JA, Montgomery JC, Wells RMG (1987) Comparative physiology of Antarctic fishes. Adv Mar Biol 24:321-388

17. Wells RMG, Macdonald JA, di Prisco G (1990) Thin-blooded Antarctic fishes: a rheological comparison of the haemoglobin-free icefishes, *Chionodraco kathleenae* and *Cryodraco antarcticus*, with a red-blooded nototheniid, *Pagothenia bernacchii*. J Fish Biol 36:595-609

18. di Prisco G, Macdonald JA, Brunori M (1992) Antarctic fishes survive exposure to carbon monoxide. Experientia 48:473-475

19. Wells RMG (1990) Hemoglobin physiology in vertebrate animals: a cautionary

approach to adaptionist thinking. In: Boutilier RG (ed) Advances in Comparative Environmental Physiology, vol 6, Springer, Berlin Heidelberg New York, pp 143-161

20. Cocca E, Ratnayake-Lecamwasam M, Parker SK, Camardella L, Ciaramella M, di Prisco G, Detrich HW III (1995) Genomic remnants of α-globin genes in the hemo-globinless Antarctic icefishes. Proc Natl Acad Sci USA 92:1817-1821

21. Cocca E, Ratnayake-Lecamwasam M, Parker SK, Camardella L, Ciaramella M, di Prisco G, Detrich HW III (1997) Do the hemoglobinless icefishes have globin genes? Comp Biochem Physiol 118A:1027-1030

22. Riggs A (1988) The Bohr effect. Ann Rev Physiol 50:181-204

23. Brittain T (1987). The Root effect. Comp Biochem Physiol 86B:473-481

24. D'Avino R, di Prisco G (1997) The hemoglobin system of Antarctic and non-Antarctic notothenioid fishes. Comp Biochem Physiol 118A:1045-1049

25. Fago A, D'Avino R, di Prisco G (1992) The hemoglobins of *Notothenia angustata*, a temperate fish belonging to a family largely endemic to the Antarctic Ocean. Eur J Biochem 210:963-970

26. D'Avino R, Caruso C, Tamburrini M, Romano M, Rutigliano B, Polverino de Laureto P, Camardella L, Carratore V, di Prisco G (1994) Molecular characterization of the functionally distinct hemoglobins of the Antarctic fish *Trematomus newnesi*. J Biol Chem 269:9675-9681

27. Tamburrini M, D'Avino R, Fago A, Carratore V, Kunzmann A, di Prisco G (1996) The unique hemoglobin system of *Pleuragramma antarcticum*, an Antarctic migratory teleost. Structure and function of the three components. J Biol Chem 271:23780-23785

28. Tamburrini M, Romano M, Carratore V, Kunzmann A, Coletta M, di Prisco G (1998) The hemoglobins of the Antarctic fishes *Artedidraco orianae* and *Pogonophryne scotti*. Amino acid sequence, lack of cooperativity and ligand binding properties. J Biol Chem 273:32452-32459

29. DeVries AL (1988) The role of antifreeze glycopeptides and peptides in the freezing avoidance of Antarctic fishes. Comp Biochem Physiol 90B:611-621

30. Cheng CC, DeVries AL (1991) The role of antifreeze glycopeptides and peptides in the freezing avoidance of cold-water fish. In: di Prisco G (ed) Life under extreme condi-tions. Biochemical adaptations. Springer, Berlin Heidelberg New York, pp 1-14

31. Cheng CC (1998) Origin and mechanism of evolution of antifreeze glycoproteins in polar fishes. In: di Prisco G, Pisano E, Clarke A (eds) Fishes of Antarctica. A biologi-cal overview. Springer, Milano Heidelberg New York, pp 311-328

32. Detrich HW III (1998) Molecular adaptation of microtubules and microtubule motors from Antarctic fish. In: di Prisco G, Pisano E, Clarke A (eds) Fishes of Antarctica. A biological overview. Springer, Milano Heidelberg New York, pp 139-149

33. Sweezey RR, Somero GN (1982) Polymerization thermodynamics and structural sta-bilities of skeletal muscle actins from vertebrates adapted to different temperatures and hydrostatic pressures. Biochemistry 21:4496-4503

34. Ciardiello MA, Camardella L, di Prisco G (1995) Glucose-6-phosphate dehydrogenase from the blood cells of two Antarctic teleosts: correlation with cold adaptation. Biochim Biophys Acta 1250:76-82

35. Ciardiello MA, Camardella L, di Prisco G (1997) Enzymes of Antarctic fishes: effect of temperature on catalysis. Cybium 21(4):443-450

36. Hochachka PW, Somero GN (1984) Biochemical adaptation. Princeton University Press, Princeton

37. Coletta M, Condò SG, Scatena R, Clementi ME, Baroni S, Sletten SN, Brix O, Giardina B (1994) Synergistic modulation by chloride and organic phosphates of hemoglobin from bear (*Ursus arctos*). J Mol Biol 236:1401-1406

38. di Prisco G, Tamburrini M (1992) The hemoglobins of marine and freshwater fish: the search for correlations with physiological adaptation. Comp Biochem Physiol 102B:661-671

39. Perutz MF (1987) Species adaptation in a protein molecule. Adv Prot Chem 36:213-244

40. Stam WT, Beintema JJ, D'Avino R, Tamburrini M, Cocca E, di Prisco G (1998) Evolutionary studies on teleost hemoglobin sequences. In: di Prisco G, Pisano E, Clarke A (eds) Fishes of Antarctica. A biological overview. Springer, Milano Heidelberg New York, pp 355-359

Adaptations for Oxygen Transport: Lessons from Fish Hemoglobins

R. E. WEBER

Hemoglobin (Hb) is a prototype of macromolecules whose functional properties are controlled by ionic effectors and a paradigm for study of the structure, function and allosteric interactions of proteins. By virtue of its well-defined roles in transporting O_2 from the respiratory surfaces to the tissues and metabolic end-products such as CO_2, protons, heat in the opposite direction, and its implication in regulating other processes in red cells, it forms an ideal model for probing the mechanisms of molecular adaptations to environmental conditions and physiological demands [16, 62].

Hemoglobin Adaptations: Why Fish?

Fish Hbs operate under much greater variation in the physicochemical conditions than Hbs of air-breathing vertebrates [10, 52, 60]. Firstly, aquatic habitats may show inordinate temporal and spatial variations in O_2 availability, changing between anoxia, hypoxia, and hyperoxia (O_2 free, O_2 poor and O_2 supersaturated), and in salinity, ionic composition, pH and temperature. Secondly, fish exhibit a greater variation in structure and efficiency of their gas exchange organs (gills, lungs, skin, swim-bladders, and pharyngeal and gut surfaces) and a lesser capacity for regulating their internal physicochemical conditions independently of environmental variables.

The O_2 binding characteristics of vertebrate blood are a product of the 'intrinsic' O_2 affinity of the Hb molecules and the interaction with allosteric effectors that modify (usually depress) O_2 affinity. Apart from protons and inorganic anions, organic phosphates play an important role in modulating O_2

Center for Respiratory Adaptation (CRA), Department of Zoophysiology, Institute of Biological Sciences, University of Aarhus, Aarhus, Denmark

G. di Prisco, B. Giardina, R.E. Weber (Eds)
Hemoglobin Function in Vertebrates.
Molecular Adaptation in Extreme and Temperate Environments
© Springer-Verlag Italia 2000

affinity. In contrast to mammals and birds where 2,3-diphosphoglycerate (DPG) and inositol pentaphosphate (IPP), respectively, are the main red cell organic effectors, lower vertebrates use adenosine triphosphate (ATP), which in some fish often occurs together with guanosine triphosphate (GTP). Potentially, a decrease in phosphate levels leading to increased O_2 affinity favours O_2 loading in the gills or lungs but hinders unloading in the tissues, and vice versa.

As in mammals, fish Hbs are tetrameric, consisting of two α and two β chains. When deoxygenated the molecules are constrained by salt bridges and hydrogen bonds in the low affinity, tense (T) three-dimensional structure that impedes O_2 binding by the hemes. Upon O_2 loading, the molecule clicks into the high affinity, relaxed (R) state. The shift is basic to homotropic interactions between the O_2 binding haeme groups (co-operativity) that is reflected in the sigmoid-shaped O_2 binding curve (Fig. 1) and increases the capacitance of blood (the amount of O_2 loaded or unloaded for a given change in O_2 tension). The organic phosphates, chloride and protons decrease O_2 affinity by binding at specific sites in the T state. O_2 unloading in the acid tissues is enhanced by the Bohr effect (inhibitory interactions between proton and O_2 binding sites). Some fish Hbs exhibit Root effects (extreme stabilisation of the T-state at low pH that prohibits full O_2 saturation even at extreme, high O_2 tensions) which play a role in the secretion of O_2 into the swim-bladder (increasing bouyancy) and eyes (raising retinal O_2 tensions) [26, 48, 57, 62].

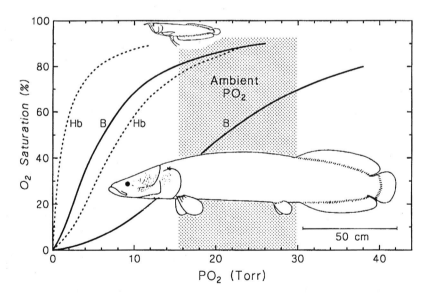

Fig. 1. O_2 equilibria of stripped (phosphate-free) Hb solutions (*Hb, dashed curves*) and whole blood (*B, solid curves*) from the surface-water breathing *Osteoglossum bicirrhosum* (*top*) and the large obligate air-breathing *Arapaima gigas* at 25°C (Hb) and 28°C (B). Modified after [29]

In contrast to mammals that have one of very few major Hb components, fish commonly have several "isoHbs" (8 to 19 in salmonids) [47] that may be polymorphic (different isoHbs occurring in different genetic strains) and often show marked functional differentiation, suggesting an in vivo division of labour [58]. In addition to electrophoretically anodic Hbs found in other vertebrates, many fish have cathodic Hbs that have higher intrinsic O_2 affinities and no (or reverse) Bohr effects (Fig. 2) and thus may function as O_2 reserves and O_2 carriers under hypoxia and activity-induced acidoses when oxygenation of anodic Hbs is compromised by Bohr and Root effects. This interpretation is supported by the high incidence of cathodic Hbs in active salmonids and catfish, and their absence in inactive benthic flatfish [58].

Following a brief discussion of the genetically coded, interspecific adaptations in Hb function, this chapter describes the intraspecific adaptations, focussing on the role of allosteric effectors and on the molecular mechanisms basic to special characteristics of fish Hbs, including the Root effect, normal and reverse Bohr effects, and the interactions with phosphates with red cell membrane constituents, such as the cytoplasmic domain of band 3 protein.

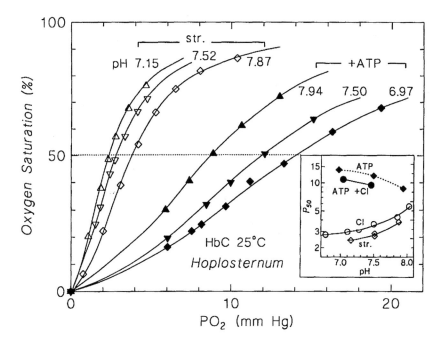

Fig. 2. O_2 equilibria of the cathodic HbC from the Amazonian catfish *Hoplosternum littorale* showing reverse Bohr effect in the stripped (phosphate-free) Hb (*open symbols*) and normal Bohr effect in the presence of ATP. *Inset:* Variation of half-saturation O_2 tension P_{50} with pH in stripped HbC in the absence of effectors (*open symbols*) and in the presence of ATP (*closed diamonds*) and ATP plus 0.10 M chloride (*closed circles*)

Blood Adaptations

Interspecific and Ontogenetic Adaptations - How Osteoglossids and Zoarcids Do It

A beautiful example of interspecific adaptation comes from the ancient osteoglossid (bonytongue) fishes, whose fossils go back to the Paleocene epoch [39]. Two of only seven known osteoglossids live side by side in hypoxic Amazon waters, namely the "pirarucu" *Arapaima gigas*, the largest known fresh water fish that obligatorily breathes air with a modified swim-bladder lung (and drowns if denied access to air), and the smaller "aruana" *Osteoglossum bicirrhosum* that is morphologically adapted to breathe surface waters. *Arapaima* exhibits markedly lower blood-O_2 affinity than *Osteoglossum* (Fig. 1), reflecting the higher O_2 tensions in air and an efficient 'lung' gas exchange. Significantly, the affinity difference persists in the "stripped" (phosphate-free) Hb solutions, showing it derives from an intrinsic O_2 affinity difference in the Hb molecules [29]. However, differential co-factor interactions may also be implicated, given that *Osteoglossum* red cells contain ATP and GTP (in a ratio of ~1 : 2.1) whereas *Arapaima* additionally has IPP [22] that depresses the O_2 affinity of fish Hb much more than ATP and GTP [29, 63, 66]. Subspecific genetic variations in fish are illustrated by small, but distinct differences in intrinsic Hb-O_2 affinities found in three different genotypes of turbot *Scophthalmus maximus* [50].

Both mechanisms (changes in intrinsic affinities and in cofactor interaction) are exploited during ontogenetic development in fish [59]. In *Arapaima* co-factor effects are indicated by the five-fold higher IPP/NTP ratios in air-breathing adults than in water-breathing juveniles [54] and from the same isoelectric points and intrinsic O_2 affinities in major Hb components from both stages (R.E. Weber and A. Val, unpublished). However, in the viviparous blenny *Zoarces viviparus*, O_2 transfer from maternal circulation to the foetuses that are retained in ovarial cavities during a 4-month gestational development is based on a higher intrinsic O_2 affinity in the foetal than in the maternal Hb [61]. An intermediate situation is illustrated in the viviparous seaperch *Embiotoca lateralis* in which maternal-foetal O_2 transfer is secured by a higher intrinsic Hb-O_2 affinity as well as by lower NTP levels in the foetal stages [21].

The fish Hb components change markedly during ontogeny, although the changes are less conspicious than in mammals that have fewer major Hb components. Generally species having both types of Hb show an increase in the ratio of cathodal to anodal components during growth [71]. The changes may be complex as illustrated in Arctic charr *Salvelinus alpinus* that expresses at least 17 Hb components during life, and 10 anodal and three cathodal components in embryonic stages, of which only five anodal and one cathodal appear in later stages [17]. The exact functional significance of these shifts remains obscure.

Adaptive Changes in Anionic Effectors

Changes in the type and concentration of red cell organic phosphates represent a main mechanism of intraspecific adaptation. These responses are rapid (detectable within 30 min and completed within days) compared to synthesis of new Hbs but slow compared to hyperventilatory responses and catecholamine interaction (see below). The action of individual exogenous and endogenous factors that elicit adaptive changes may be inter-related. For example, increases in activity, temperature and environmental pollutants may modify blood-O_2 affinity via decreases in tissue O_2 tension.

Ambient hypoxia is the most intensively studied factor. In eels, blood O_2 affinity increases under hypoxia due to decreased NTP levels that safeguard O_2 loading in the gills and the venous O_2 reserve [65, 72, 73]. Similar responses have been observed in a range of fish [42, 43, 60]. However, the exact mechanisms by which red cell ATP and GTP levels are adjusted remain unknown [43]. In tench exposed to combined hypoxia and hypercapnia, the potentially deleterious pH decrease is countered by a large increase in plasma HCO_3^- levels [23]. Apart from its direct allosteric effect, the fall in red cell NTP concentration raises blood O_2 affinity indirectly via a perturbation of the Donnan distribution of protons across the red cell membranes that raises cellular pH [74]. Amongst marine fish, the capacity for co-factor modulation as judged by NTP/Hb ratios is high in teleosts, intermediate in sharks and low in rays [70].

Where GTP as well as ATP are present, GTP plays a greater modulator role due to its greater decreases in concentration under hypoxia [23, 63, 66], its greater allosteric effect on O_2 affinity at the same NTP/Hb ratio, and its lesser inhibition by intracellular Mg^{2+} ions [36, 63, 65], which may free ATP for more primitive energy-related commitments [57]. The distribution of GTP in fish defies strict characterization in terms of O_2 availability and phylogeny. However, GTP/ATP ratios appear to be low in salmonids and marine elasmobranchs that frequent O_2 rich waters and in pleuronectiformes (flatfish) that are inactive, but high in anguilliformes (eels) that tolerate hypoxic conditions [32, 33].

Internal hypoxia may also result from anaemia or from obstruction of gas exchange or blood circulation. In trout, blood volume reduction increases erythropoiesis and raises the erythrocytic NTP concentration and the NTP/Hb ratio [31] and lowers the erythrocytic O_2 affinity [55]. However, exposing tench *Tinca tinca* to aluminium-containing water, which clogs the gills with mucus and decreases aortic PO_2, lowers NTP and GTP/Hb levels [23, 24]). The difference may relate to the fact that trout frequents well-aerated waters where a decreased O_2 affinity will not significantly jeopardize O_2 loading, whereas in tench that commonly faces environmental hypoxia it is important to safeguard O_2 loading.

Air breathing and estivation induced by water hypoxia or drought similarly evoke adaptive changes in blood O_2 binding. In the catfish, *Hypostomus* and

Pterygoplicthys, air breathing in hypoxic water results in decreased erythrocytic GTP levels, which predictably raise O_2 affinity and favour exploitation of O_2 in the swallowed air [66]. In the African lungfish *Protopterus amphibius*, drought-provoked estivation increases blood O_2 affinity via decreases in NTP (chiefly GTP) levels. Apart from favouring O_2 loading in compensation for decreased pulmonary ventilation [28], the affinity increase may curtail tissue unloading and thus be instrumental in the down-regulation of organismic metabolism during the extended periods of suspended animation [57].

Activity and blood pH changes similarly evoke direct and compensatory effects on blood O_2 affinity. In carp, where falling blood pH is associated with increased steady state NTP/Hb ratios, the resultant decreases in O_2 affinity and arterial Hb O_2 saturation increases proton binding to Hb, thus raising blood pH via negative feedback control [25].

During exercise, the Hb levels in fish blood may increase rapidly (3-5 min) due to recruitment of red cells from splenic stores [75]. The NTP concentrations in spleen-stored trout red cells are one third higher than those in circulation [69], suggesting that the liberation of stored cells may favour tissue O_2 unloading. However, no effects on blood O_2 affinity and ATP/Hb ratio are evident in trout after 200 days of exercise [8].

Molecular Mechanisms of Allosteric Effectors

Phosphate Interactions: Implications of His-NA2(β)

In humans, DPG binds to deoxyHb at seven sites, namely Val-NA1(1) of one, and His NA2 (2), Lys EF6(82), and His-H21(143) of both β-chains [46, 60]. In fish Hb a binding site stereochemically complementary to strain-free NTP results from the presence of Glu or Asp at NA2($\beta2$) and Arg at H21(β) [48] (Table 1). Modelling attributes the greater effect of GTP than ATP on O_2 affinity of carp Hb to the formation of hydrogen bonds with Val-NA1 of both β-chains, compared to one chain in the case of ATP [19]. The difference in the effects of GTP and ATP on the O_2 affinity of carp Hb accords with the free energy of a single bond contributing to the stability of the T structure [67, 63]. Curiously, however, ATP and GTP effects are similar in the Hb of a few Amazonian teleosts [53, 60].

Apart from NTP, the erythrocytes of a few, phylogenetically unrelated fish contain significant DPG concentrations [52, 68]. In the facultative air breathing Amazonian armoured catfish *Hoplosternum littorale*, DPG levels approximate those of ATP and GTP and increase with increasing ambient temperature [52]. Although having 'mammalian' His (instead of 'piscine' Glu) at NA2(β), cathodic *Hoplosternum* HbC exhibits a much lower sensitivity to DPG than to ATP or GTP [68]. In life, however, NTP effects are inhibited more by complex formation with divalent cations, given that the Mg^{2+}-ATP stability constant exceeds that of DPG by an order of magnitude [5].

What then, if any, might be the significance of His at NA2(β)? The lower sensitivity of *Hoplosternum* HbC to DPG than to NTP suggests that it does not increase Hb's potential for DPG modulation. Amongst ectothermic vertebrates, it is also encountered in the lungfish *Lepidosiren paradoxa*, the sharks *Squalus acanthias* and *Heterodontus portusjacksoni*, and in tadpoles of the frog *Rana catesbeiana* and the toad *Xenopus laevis* [30], suggesting that it may be a primitive vertebrate character or related to air-breathing. An alternative interpretation (see below) is that it is involved in the interaction of Hb with cd-B3 (the cytoplasmic domain of the red cell membrane protein band 3).

Bohr Effects

In human Hb, the proton binding responsible for the normal ('alkaline') Bohr effect occurs mainly at the N-terminal Val-NA1(1) residues of the α-chains and the C-terminal His-HC3(146) residues of the β-chains. Given that Val-NAα1 is acetylated in fish Hb, the absence of a normal Bohr effect in cathodic fish Hb correlates with substitution of His-HC3(β146) by Phe (Table 1).

Table 1. Some functionally significant β-chain residues in anodic and cathodic fish Hbs. D, H, Cl indicate DPG, proton and chloride binding sites in human Hb

Helical position	NA1(1)	NA2(2)	EF6(82)	F9	H21(143)	HC3(146)	Reference
	D	D H	D H Cl		D H	H	
Anodic Hbs							
Human HbA	Val	His	Lys	Cys	His	His	[9]
Carp *Cyprinus carpio* (R)	Val	Glu	Lys	Ser	Arg	His	[30]
Trout *Oncorhynchus mykiss* Hb IV (R)	Val	Asp	Lys	Ser	Arg	His	[30]
Eel *Anguilla anguilla* HbA (R)	Val	Glu	Lys	Ser	Arg	His	[12]
Cathodic fish Hbs							
Trout *Oncorhynchus mykiss* Hb I	Val	Glu	Leu	Ala	Ser	Phe	[30]
Eel *Anguilla anguilla* HbC (rB)	Val	Glu	Lys	Asn	Lys	Phe	[11]
Catfish *Hoplosternum litorale* Hb C (rB)	Val	His	Leu	Ala	Ser	Phe	[68]

R, Hbs expressing Root effects; rB, Hbs with reverse Bohr effects in the absence of phosphates

In contrast to the intensively studied trout Hb I that exhibits no Bohr effect, stripped cathodic Hbs of non-salmonid fish (eel *Anguilla*, catfish *Hoplosternum*, *Pterygoplichthys*, and characin *Mylossoma*) [11, 15, 18, 37, 64] show pronounced reverse Bohr effects that are obliterated or become normal in the presence of organic phosphates (Fig. 2). Although the reverse Bohr effect will thus be suppressed in life, its presence in stripped Hb may be essential to obtain small in vivo Bohr effects in phosphate-sensitive Hbs, since phosphate effects increase with falling pH and otherwise would induce inordinately large normal Bohr effects (see Fig. 2).

The reverse Bohr effect of the cathodic Hbs implies oxygenation-linked proton binding, rather than proton release as in anodic Hbs with normal Bohr effect, and thus indicates either the presence of alternative ionizable groups or radical modification of alkaline Bohr groups. The His-H21 (β143) residue that is considered to be implicated in the reverse Bohr effect seen in mammalian Hb at acid pH is substituted in cathodic fish Hbs (Table 1), indicating a different allosteric mechanism.

The view that the phosphate binding sites in the cavity between the β-chains are implicated in expression of the reverse Bohr effect [11] is based on repulsion between these positively charged sites that may destabilize the T state relative to the R state [2], whereby Cl may lower affinity by neutralizing these charges [49]. The close packing of the positive side chains in this cavity may lower the pK_a values of Val-NA1β and possibly even the Lys-EF6β residues making them acid Bohr groups [11]. This hypothesis is consistent with the loss of the reverse Bohr effect in the presence of phosphates that neutralize the positively charged sites (Fig. 2).

Root Effect

A diagnostic feature of the Root effect is the large decrease in Hb O_2 affinity and co-operativity at low pH, reflecting extreme stabilization of the T state that prohibits full O_2 saturation even at O_2 tensions exceeding 100 atm [3]. The loss of co-operativity at low pH appears to be related to large differences in affinity of the α and β subunits, as indicated by biphasic CO association kinetics of the tetrameric molecules [44].

Perutz and Brunori [48] proposed that a constellation of polar residues that includes hydrophilic Ser F9(β), which allows formation of two additional H bonds in the T state, is needed to produce a Root effect (see Table 1). However, human Hb mutated at F9(β) by site-directed mutagenesis, namely Hb Nympheas (Cys-F9 $\rightarrow$Ser) and Hb Daphne (Cys-F9 $\rightarrow$Ser + His-H21$\rightarrow$Arg) [41] and Hb Sandra (Cys-F9 $\rightarrow$Ser + Asp-FG1$\rightarrow$Glu) [35] failed to show a Root effect. An alternative mechanism proposed on the basis of work with spot (*Leiostomus xanthurus*) Hb [40] explains the Root effect in terms of protonation of the β-chain NA1 and HC3 residues that occurs within positively charged clusters assembled across the β1/β2 interface and destabilizes the R state at low pH.

The fact that Root effect Hbs are frozen in the T state at low pH where non-Root effect Hbs undergo the co-operative transition indicates a crucial role for the α_1/β_2 contact in expression of the Root effect. Comparison between Root effect and non-Root effect Hbs indicates that the residues at positions C3, C6 and CD2 of the α-chain and FG4 of the β-chain may be crucial in the expression of the Root effect [12, 26]. The role of these residues is currently under investigation in our laboratory using cloned anodic and cathodic eel Hbs.

Allosteric Control Mechanisms

As seen with DPG and human Hb, ATP and GTP decrease the O_2 affinity of anodic fish Hbs by lowering K_T (the affinity constants of the Hb in the deoxygenated state) without significantly affecting that (K_R) of the oxygenated state (Fig. 3). The phosphates thus increase co-operativity, and the free energy for heme-heme interaction ΔG (whose magnitude is reflected in the distance between the upper and lower asymptotes of extended Hill plots; see Fig. 3) and thus the capacitance of the blood for O_2 transport, GTP exerting the greater effect (Fig. 3).

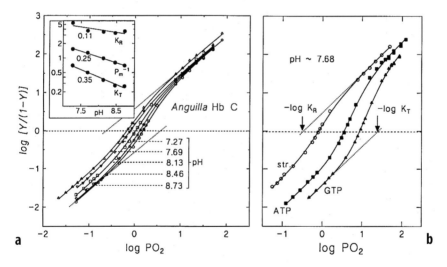

Fig. 3a,b. Extended Hill plots for cathodic eel *Anguilla anguilla* HbC at 15°C, at the indicated pH values (a), and in the absence (stripped) and presence of ATP and GTP (NTP/Hb tetramer ratios of approximately 20) (b). As indicated, intersections of the lower and upper asymptotes with the x-axis at log $[Y/(1-Y)] = 0$ (where Y is the fractional 0_2 saturation of the Hb) indicate the K_T and K_R values (O_2 binding constants of the Hb in the T and R states, respectively). As illustrated (a) the reverse Bohr effect is mainly due to an increase in K_T (in contrast to human and other anodic Hbs where increasing proton concentrations decrease K_T) and (b) organic phosphates modulate O_2 affinity mainly by decreasing K_T, GTP exerting a greater effect than ATP. *Inset*: Linear regressions of K_T, P_m^{-1} and K_R plotted against pH showing Bohr effects for binding the first, median and last O_2 molecules, respectively, to the Hb molecules. Modified after [14]

A homologous control mechanism (decrease in K_T) underlies the effect of increased proton concentrations (the normal Bohr effect) in human Hb and in anodic fish Hb in the (upper) pH range where the Root effect is absent [7, 20, 67]. Deviations from this pattern apply to Root effect Hbs at low pH, where the loss of co-operativity is explained by large decreases in K_R [67] - or by the molecules remaining frozen in the T state upon O_2 binding [4, 12] - and to cathodic Hbs, where the reverse Bohr effects correlate with *increased* K_T values that decrease ΔG as pH falls [14, 68] (Fig. 3). Cathodic Hbs thus undergo a decrease in the bond energies that constrain the Hb in the T state as pH decreases, in contrast to human Hb where additional bonds are formed [45].

Band 3 and Red Cell-Mediated Hb Interactions

The reaction of Hb with the cytoplasmic domain of band 3 (cd-B3) that constitutes a major part (25%) of the erythrocytic membrane proteins [51] implies that Hb may function as transducer for a variety of cellular processes, since band 3 protein is responsible for HCO_3^-/Cl^- exchange across the red cell membrane, and its cytoplasmic domain also binds to several glycolytic enzymes (such as aldolase, phosphofructokinase, glyceraldehyde-3-phosphate dehydrogenase and lactate dehydrogenase) [16, 34]. Cd-B3 and synthetic peptides of the corresponding structure bind between the β-chains of human Hb in competition with DPG and analogously lowers Hb O_2 affinity [6, 38, 56]. The emerging inference is that the Hb-band 3 interaction may link red cell glycolysis with O_2 transport . Thus, at high O_2 tensions the low affinity of oxyHb for cdB3 would favour formation of cdB3-glycolytic enzyme complexes that inhibit the glycolytic pathway, whereas at low O_2 levels the high affinity of deoxyHb for cdB3 frees the enzymes for activating the glycolytic pathway. The dependence of a catecholamine-induced swelling of trout red cells on (CO or O_2) ligation of the Hb suggests a transducer role for fish Hb [16].

Curiously, synthetic peptides corresponding to the first 10 and 20 amino acid residues of the N-terminal fragment of trout band 3 protein have no tangible effect on O_2 affinity of trout Hb IV but markedly decreases that of human Hb [27]. The lack of oxylabile interaction with trout Hb IV stands confirmed for cathodic trout Hb I and trout Hbs II and III (Fig. 4).

In contrast to trout Hb, a recent study [68] shows that trout cd-B3 peptides interact with *Hoplosternum* HbC that has positively charged His at NA2(β2). This accords with the view that the lack of interaction with trout Hb IV is related to the presence of negatively charged Asp-NA(2)(β2) [27] and imparts novel significance to the occurrence of His-NA2(β2) in some amphibious and other ectotherms (see above), namely that it may endow Hbs with a transducer role in regulating glycolysis in an O_2-dependent manner.

An alternative rapid mechanism for red cell O_2 affinity regulation is adrenergic stimulation, which has recently been reviewed [42, 43]. Briefly, catecholamine binding to red cell adrenergic receptors activates adenylate cyclase,

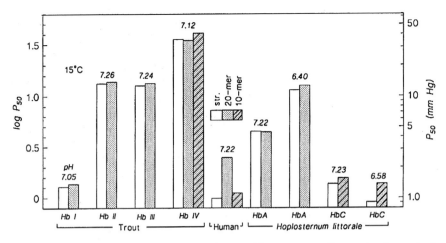

Fig. 4. Effects of 10 and 20 amino acid residue peptides corresponding to the N-terminal part of the cytoplasmic domain of trout band 3 protein (cd-B3) on the half-saturation O_2 tensions (P_{50}) of trout Hbs I, II, III and IV, human Hb, and *Hoplosternum littorale* HbA and HbC measured at 15°C and the indicated pH values. Data for human Hb are from [27]. Hb concentrations, 0.08 mM tetramer (fish Hbs) and 0.09-0.16 mM tetramer (human Hb). Peptide/Hb (tetramer) molar ratios ~5

forming cAMP that activates a sodium(in)/protons(out) exchanger. This alkalinizes the intracellular environment that raises Hb O_2 affinity via the Bohr effect and causes in a temporary anion disequilibrium that induces a HCO_3^- (out)/Cl^-(in) exchange. The accumulation of Na^+ and Cl^- ions causes water imbibition and cellular swelling that dilutes Hb and phosphate concentrations, resulting in a further increase in O_2 affinity.

As previously observed in crocodilian Hb [1], Hb of the hagfish *Myxine glutinosa* shows O_2-linked binding of bicarbonate that decreases O_2 affinity. Given the virtual absence of band 3 protein in this species, this effect may enhance O_2 unloading to the tissues and facilitate the hydration of CO_2 and its transport as intraerythrocytic HCO_3^- [13].

Concluding Remarks

The functional and molecular properties of fish Hb described here are part of a symphony of adaptations manifested at the molecular, cellular and systemic levels of biological organization (including hyperventilatory and cardiovascular responses that were not dealt with in this chapter). In concert, these adaptations safeguard blood O_2 transport in response to specific ambient conditions and endogenous factors (such as breathing mode), increasing the tolerance of animals to environmental vagaries and extending their range of habitable environments.

References

1. Bauer C, Forster M, Gros G, Mosca A, Perrella M, Rollema HS, Vogel D (1981) Analysis of bicarbonate binding to crocodilian hemoglobin. J Biol Chem 256:8429-8435

2. Bonaventura C, Bonaventura J (1978) Anionic control of hemoglobin function. In: Caughey WS (ed) Biochemical and clinical aspects of hemoglobin abnormalities. Academic, New York, pp 647-663

3. Brittain T (1987) The Root effect. Comp Biochem Physiol 86B:473-481

4. Brunori M, Coletta M, Giardina B, Wyman J (1978) A macromolecular transducer as illustrated by trout hemoglobin IV. Proc Natl Acad Sci USA 75:4310-4312

5. Bunn HF, Ransil BJ, Chao A (1971) The interaction between erythrocyte organic phosphates, magnesium ion, and hemoglobin. J Biol Chem 246:5273-5279

6. Chétrite G, Cassoly R, Chetrite G (1985) Affinity of hemoglobin for the cytoplasmic fragment of human erythrocyte membrane band 3. Equilibrium measurements at physiological pH using matrix-bound proteins: the effects of ionic strength, deoxygenation and of 2,3-diphosphoglycerate. J Mol Biol 185:639-644

7. Chien JCW, Mayo KH (1980) Carp hemoglobin. I. Precise oxygen equilibrium and analysis according to the models of Adair and of Monod, Wyman, and Changeux. J Biol Chem 255:9790-9799

8. Davie PS, Wells RMG, Tetens V (1986) Effects of sustained swimming on rainbow trout muscle structure, blood oxygen transport, and lactate dehydrogenase isozymes: evidence for increased aerobic capacity of white muscle. J Exp Zool 237:159-171

9. Davis BJ (1991) Developmental changes in the blood oxygen transport system of Kemp's ridley sea turtle, Lepidochelys kempi. Can J Zool 69:2660-2666

10. di Prisco G, Tamburrini M (1992) The hemoglobins of marine and freshwater fish: The search for correlations with physiological adaptation. Comp Biochem Physiol [B] 102:661-671

11. Fago A, Carratore V, di Prisco G, Feuerlein RJ, Sottrup-Jensen L, Weber RE (1995) The cathodic hemoglobin of Anguilla anguilla. Amino acid sequence and oxygen equilibria of a reverse Bohr effect hemoglobin with high oxygen affinity and high phosphate sensitivity. J Biol Chem 270:18897-18902

12. Fago A, Bendixen E, Malte H, Weber RE (1997) The anodic hemoglobin of Anguilla anguilla. Molecular base for allosteric effects in a Root-effect hemoglobin. J Biol Chem 272:15628-15635

13. Fago A, Malte H, Dohn N (1999) Bicarbonate binding to hemoglobin links oxygen and carbon dioxide transport in hagfish. Respir Physiol 115:309-315

14. Feuerlein RJ, Weber RE (1996) Oxygen equilibria of cathodic eel hemoglobin analysed in terms of the MWC model and Adair's successive oxygenation theory. J Comp Physiol 165:597-606

15. Garlick RL, Bunn HF, Fyhn HJ, Fyhn UEH, Martin JP, Noble RW, Powers D (1979) Functional studies on the separated hemoglobin components of an air-breathing catfish, Hoplosternum littorate (Hancock). Comp Biochem Physiol 62A:219-226

16. Giardina B, Messana I, Scatena R, Castagnola M (1995) The multiple functions of hemoglobin. Crit Rev Biochem Mol Biol 30:165-196

17. Giles MA, Rystephanuk DM (1989) Ontogenetic variation in the multiple hemoglobins of Arctic charr, Salvelinus alpinus. Can J Fish Aquat Sci 46:804-809

18. Gillen RG, Riggs A (1973) Structure and function of the isolated hemoglobins of the American eel, Anguilla rostrata. J Biol Chem 248:1961-1969

19. Gronenborn AM, Clore GM, Brunori M, Giardina B, Falcioni G, Perutz MF (1984)

Stereochemistry of ATP and GTP bound to fish haemoglobins. A transferred nuclear overhauser enhancement, 31P-Nuclear Magnetic Resonance, oxygen equilibrium and molecular modelling study. J Mol Biol 178:731-742

20. Imai K (1982) Allosteric effects in haemoglobin. Cambridge University Press, Cambridge

21. Ingermann RL, Terwilliger RC (1981) Intraerythrocytic organic phosphates of fetal and adult seaperch (Embiotoca lateralis): their role in maternal-fetal oxygen transport. J Comp Physiol [B] 144:253-259

22. Isaacks RE, Kim HD, Harkness DR (1978) Relationship between phosphorylated metabolic intermediates and whole blood oxygen affinity in some air-breathing and water-breathing teleosts. Can J Zool 56:887-890

23. Jensen FB, Weber RE (1982) Respiratory properties of tench blood and hemoglobin. Adaptation to hypoxic-hypercapnic water. Molec Physiol 2:235-250

24. Jensen FB, Weber RE (1987) Internal hypoxia-hypercapnia in tench exposed to aluminium in acid water: effects on blood gas transport, acid-base status and electrolyte composition in arterial blood. J Exp Biol 127:427-442

25. Jensen FB, Andersen NA, Heisler N (1990) Interrelationships between red cell nucleoside triphosphate content, and blood pH, O2-tension and haemoglobin concentration in carp, Cyprinus carpio. Fish Physiol Biochem 8:459-464

26. Jensen FB, Fago A, Weber RE (1998) Hemoglobin structure and function. In: Perry SF, Tufts BL (eds) Fish respiration. Academic, San Diego, pp 1-40

27. Jensen FB, Jakobsen MH, Weber RE (1998) Interaction between haemoglobin and synthetic peptides of the N-terminal cytoplasmic Interaction between haemoglobin and synthetic peptides of the N-terminal cytoplasmic fragment of trout band 3 (AE1) protein. J Exp Biol 201:2685-2690

28. Johansen K, Lykkeboe G, Weber RE, Maloiy GMO (1976) Respiratory properties of blood in awake and estivating lungfish, Protopterus amphibius. Respir Physiol 27:335-345

29. Johansen K, Mangum CP, Weber RE (1978) Reduced blood O2 affinity associated with air breathing in osteoglossid fishes. Can J Zool 56:891-897

30. Kleinschmidt T, Sgouros JG (1987) Hemoglobin sequences. Biol Chem Hoppe-Seyler 368:579-615

31. Lane HC, Rolfe AE, Nelson JR (1981) Changes in the nucleoside triphosphate/haemoglobin and nucleoside triphosphate/red cell ratios of rainbow trout, Salmo gairdneri Richardson, subjected to prolonged starvation and bleeding. J Fish Biol 18:661-668

32. Leray C (1979) Patterns of purine necleotides in fish erythrocytes. Comp Biochem Physiol [B] 64B:77-82

33. Leray C (1982) Patterns of purine nucleotides in North Sea fish erythrocytes. Comp Biochem Physiol 71B:77-81

34. Low PS (1986) Structure and function of the cytoplasmic domain of band 3: center of erythrocyte membrane-peripheral protein interactions. Biochim Biophys Acta 864:145-167

35. Luisi BF, Nagai K, Perutz M, Perutz MF (1987) X-ray crystallographic and functional studies of human haemoglobin mutants produced in Escherichia coli. Acta Haematol 78:85-89

36. Lykkeboe G, Johansen K, Maloiy GMO (1975) Functional properties of hemoglobins in the teleost Tilapia grahami. J Comp Physiol 104:1-11

37. Martin JP, Bonaventura J, Brunori M, Fyhn HJ, Fyhn UEH, Garlick RL, Powers DA, Wilson MT (1979) The isolation and characterization of the hemoglobin components of Mylossoma sp., an Amazonian teleost. Comp Biochem Physiol 62A:155-162

38. Messana I, Orlando M, Cassiano L, Pennacchietti L, Zuppi C, Castagnola M, Giardina B (1996) Human erythrocyte metabolism is modulated by the O2-linked transition of hemoglobin. FEBS Lett 390:25-28

39. Moyle PB, Cech JJ Jr (1996) Fishes: an introduction to ichthyology, 3rd edn. Prentice Hall, Upper Saddle River

40. Mylvaganam SE, Bonaventura C, Bonaventura J, Getzoff ED (1996) Structural basis for the Root effect in haemoglobin. Nature Struct Biol 3:275-283

41. Nagai K, Perutz MF, Poyart C (1985) Oxygen binding properties of human mutant hemoglobins synthesized in Escherichia coli. Proc Natl Acad Sci USA 82:7252-7255

42. Nikinmaa M (1992) Membrane transport and control of hemoglobin-oxygen affinity in nucleated erythrocytes. Physiol Rev 72:301-321

43. Nikinmaa M, Salama A (1998) Oxygen transport in fish. In: Perry SF, Tufts BL (eds) Fish respiration. Academic, San Diego, pp 141-184

44. Noble RW, Kwiatkowski LD, De Young A, Davis BJ, Haedrich RL, Tam LT, Riggs AF, Tam LT (1986) Functional properties of hemoglobins from deep-sea fish: correlations with depth distribution and presence of a swimbladder. Biochim Biophys Acta 870:552-563

45. Perutz MF (1970) Stereochemistry of cooperative effects in haemoglobin. Haem-haem interaction and the problem of allostery. Nature 228:726-734

46. Perutz MF (1983) Species adaptation in a protein molecule. Mol Biol Evol 1(1):1-28

47. Perutz MF (1986) A bacterial haemoglobin. Nature 322:405

48. Perutz MF, Brunori M (1982) Stereochemistry of cooperative effects in fish and amphibian haemoglobins. Nature 299(5882):421-426

49. Perutz MF, Fermi G, Poyart C, Pagnier J, Kister J (1993) A novel allosteric mechanism in haemoglobin. Structure of bovine deoxyhaemoglobin, absence of specific chloride-binding sites and origin of the chloride-linked Bohr effect in bovine and human haemoglobin. J Mol Biol 233:536-545

50. Samuelsen EN, Imsland AK, Brix O (1999) Oxygen binding properties of three different hemoglobin genotypes in turbot (Scophthalmus maximus Rafinesque): effect of temperature and pH. Fish Physiol Biochem 20:135-141

51. Sayare M, Fikiet M (1981) Cross-linking of hemoglobin to the cytoplasmic surface of human erythrocyte membranes. Identification of band 3 as a site for hemoglobin binding in Cu2+-o-phenanthroline catalyzed cross-linking. J Biol Chem 256:13152-13158

52. Val AL (1993) Adaptations of fishes to extreme conditions in fresh waters. In: Bicudo JEPW (ed) The vertebrate gas transport cascade. Adaptations to environment and mode of life. CRC Press, Boca Raton, pp 43-53

53. Val AL, Schwantes AR, Almeida-Val VMF (1986) Biological aspects of Amazonian fishes VI. Hemoglobins and whole blood properties of Semaprochilodus species (Prochilodontidae) at two phases of migration. Comp Biochem Physiol 83B:659-667

54. Val AL, Affonso EG, Souza RHS, Almeida-Val VMF, Moura MAF (1992) Inositol pentaphosphate in the erythrocytes of an Amazonian fish, the pirarucu (Arapaima gigas). Can J Zool 70:852-855

55. Vorger P, Ristori M-T (1985) Effects of experimental anemia on the ATP content and the oxygen affinity of the blood in the rainbow trout (Salmo gairdnerii). Comp Biochem Physiol A 82A:221-224

56. Walder JA, Chatterjee R, Steck TL, Low PS, Musso GF, Kaiser ET, Rogers PH, Arnone A (1984) The interaction of hemoglobin with the cytoplasmic domain of Band 3 of the human erythrocyte membrane. J Biol Chem 259:10238-10246

57. Weber RE (1982) Intraspecific adaptation of hemoglobin function in fish to oxygen availability. In: Addink ADF, Spronk N (eds) Exogenous and endogenous influences on metabolic and neural control, vol 1. Pergamon, Oxford, pp 87-102

58. Weber RE (1990) Functional significance and structural basis of multiple hemoglobins with special reference to ectothermic vertebrates. In: Truchot JP, Lahlou B (eds) Animal nutrition and transport processes. 2. Transport, respiration and excretion: comparative and environmental aspects. Comparative physiology, vol 6, Basel, Karger, pp 58-75

59. Weber RE (1994) Hemoglobin-based O2 transfer in viviparous animals. Isr J Zool 40:541-550

60. Weber RE (1996) Hemoglobin adaptations in Amazonian and temperate fish with special reference to hypoxia, allosteric effectors and functional heterogeneity. In: Val AL, Almeida-Val VMF, Randall DJ (eds) Physiology and biochemistry of the fishes of the Amazon. INPA, Brazil, pp 75-90

61. Weber RE, Hartvig M (1984) Specific fetal hemoglobin underlies the fetal-maternal shift in blood oxygen affinity in a viviparous teleost. Molec Physiol 6:27-32

62. Weber RE, Jensen FB (1988) Functional adaptations in hemoglobins from ectothermic vertebrates. Annu Rev Physiol 50:161-179

63. Weber RE, Lykkeboe G (1978) Respiratory adaptations in carp blood. Influences of hypoxia, red cell organic phosphates, divalent cations and CO2 on hemoglobin-oxygen affinity. J Comp Physiol 128B:127-137

64. Weber RE, Wood SC (1979) Effects of erythrocytic nucleoside triphosphates on oxygen equilibria of composite and fractionated hemoglobins from the facultative air-breathing Amazonian catfish, Hypostomus and Pterygoplichthys. Comp Biochem Physiol 62A:179-183

65. Weber RE, Lykkeboe G, Johansen K (1976) Physiological properties of eel haemoglobin: Hypoxic acclimation, phosphate effects and multiplicity. J Exp Biol 64:75-88

66. Weber RE, Wood SC, Davis BJ (1979) Acclimation to hypoxic water in facultative air-breathing fish: Blood oxygen affinity and allosteric effectors. Comp Biochem Physiol 62A:125-129

67. Weber RE, Jensen FB, Cox RP (1987) Analysis of teleost hemoglobin by Adair and Monod-Wyman-Changeux models. Effects of nucleoside triphosphates and pH on oxygenation of tench hemoglobin. J Comp Physiol 157B:145-152

68. Weber RE, Fago A, Val AL, Bang A, Van Hauwaert M-L, Dewilde S, Zal F, Moens L (2000) Isohemoglobin differentiation in the bimodal-breathing Amazon catfish Hoplosternum littorale. J Biol Chem, in press

69. Wells RMG, Weber RE (1990) The spleen in hypoxic and exercised rainbow trout. J Exp Biol 150:461-466

70. Wilhelm Filho D, Marcon JL, Caprario FX, Correa Nollis A (1992) Erythrocytic nucleoside triphosphates in marine fish. Comp Biochem Physiol A 102A:323-331

71. Wilkins NP (1985) Ontogeny and evolution of salmonid hemoglobins. Int Rev Cytol 94:269-298

72. Wood SC, Johansen K (1972) Adaptation to hypoxia by increased HbO2 affinity and decreased red cell ATP concentration. Nat New Biol 237:278-279

73. Wood SC, Johansen K (1973) Blood oxygen transport and acid-base balance in eels during hypoxia. Am J Physiol 225:849-851

74. Wood SC, Johansen K (1973) Organic phosphate metabolism in nucleated red cells: influence of hypoxia on eel HbO2 affinity. Neth J Sea Res 7:328-338

75. Yamamoto KI (1988) Contraction of spleen in exercised freshwater teleost. Comp Biochem Physiol A 89:65-66

Recent Evolution of the Hemoglobinless Condition of the Antarctic Icefishes

H.W. Detrich, III

Introduction

Certainly one of the most unusual "adaptations" of vertebrates is the loss of erythrocytes and the oxygen transport protein hemoglobin [1] by the Antarctic icefishes (family Channichthyidae, suborder Nototheniodei). Lacking an oxygen transporter, the icefishes nonetheless maintain normal metabolic function by delivering oxygen to their tissues in physical solution in their "colorless" or "white" blood. In the chronically cold (-1.86 to +1 C) and oxygen-rich environment experienced by these psychrophilic organisms, reduction of the hematocrit to near zero appears to have been selectively advantageous because it significantly diminishes the energetic cost associated with circulation of a highly viscous, corpuscular blood fluid [2-5]. Hematocrit, mean cellular hemoglobin concentration, and hemoglobin chain multiplicity all decrease with increasing phylogenetic divergence among the red-blooded Antarctic notothenioid fishes [6], and the Bathydraconidae (the sister group to the channichthyids) approach the hematological extreme displayed by the white-blooded icefishes. Nevertheless, the development in icefishes of compensatory physiological and circulatory adaptations that reduce tissue oxygen demand and enhance oxygen delivery (e.g., modest suppression of metabolic rates, enhanced gas exchange by large, well-perfused gills and a scaleless skin, and large increases in cardiac output and blood volume) argues that loss of hemoglobin and erythrocytes was probably maladaptive under conditions of physiological stress. Therefore, the most plausible evolutionary scenario is that the phylogenetic trend to reduced hematocrits and decreased hemoglobin synthesis in notothenioid fishes developed concurrently with enhancements to their respiratory and circulatory systems, leading ultimately to the acorpuscular, hemoglobinless condition of the icefishes.

Department of Biology, Northeastern University, Boston, MA 02115, USA

G. di Prisco, B. Giardina, R.E. Weber (Eds)
Hemoglobin Function in Vertebrates.
Molecular Adaptation in Extreme and Temperate Environments
© Springer-Verlag Italia 2000

The major adult α1- and β-globin genes of red-blooded notothenioids, like those of many temperate fishes [7-9], are tightly linked in 5' to 5' orientation (D. Lau, A. Saeed, S.K. Parker, and H.W. Detrich, III, unpublished results). What, then, has been the evolutionary fate of the adult globin genes of the non-globin-expressing icefishes? Here I synthesize work from my laboratory that demonstrates that icefish species belonging to both primitive and advanced genera have lost the gene that encodes adult β globin but retain in their genomes inactive remnants of the adult notothenioid α1-globin gene [10, 11]. Thus, the hemoglobinless phenotype of icefishes appears to be a primitive character that was established in the ancestral channichthyid by simultaneous deletion of the gene encoding β globin and part of the linked α1-globin gene prior to diversification within the clade [11]. The residual α1-globin gene of the icefishes appears to be mutating randomly. Using transversion substitutions, one can estimate that these nonfunctional nuclear gene fragments are diverging at the rate of 0.12%-0.33% per million years. This low divergence rate, which is among the smallest observed in poikilotherms, is consistent with their low, specific metabolic rates [2, 12].

The Adult Globin Gene Complex of a Red-Blooded Antarctic Nototheniid Fish

To establish a comparative framework for analysis of icefish globin genes, we isolated and characterized in my laboratory the adult α1/β-globin gene complex from the closely related, but hemoglobin-expressing, notothenioid *Notothenia coriiceps* (family Nototheniidae; D. Lau, A. Saeed, S.K. Parker, and H.W. Detrich, III, unpublished results). Figure 1 shows that the complex was composed of the α1-globin gene linked in head-to-head (5' to 5') orientation with the gene for β globin, with 4.3 kilobases (kb) of intergenic DNA separating the start codons of the two genes. The α1- and the β-globin genes were each composed of three exons separated by two introns, and the positions of their introns conformed to the vertebrate norms for globin genes [13-15]. That this complex constitutes the functional globin locus of adult *N. coriiceps* was demonstrated by: (1) the exact match of the coding sequences and the 5'- and 3'-untranslated regions of the α1- and β-globin genes of *N. coriiceps* to the corresponding regions of the adult globin cDNAs [10,11] of this species; and (2) by the identity of the primary sequences of the encoded α1 and β globins to those obtained by automated Edman degradation of the major adult globin polypeptide chains [16, 17]. Furthermore, the *N. coriiceps* globin genes lacked the structural features characteristic of typical globin pseudogenes [18, 19] or processed pseudogenes [20].

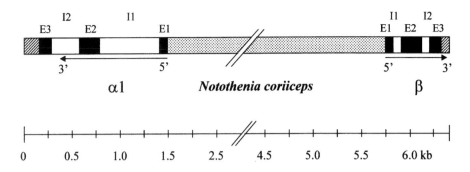

Fig. 1. Organization of the adult globin gene complex of the Antarctic rockcod *N. coriiceps*. The α1- and β-globin genes are linked in 5' to 5' orientation. The exons (*E1-E3*) and introns (*I1, I2*) of the globin genes are represented by *filled* and *open rectangles*, respectively. The intergenic region, which is defined as the 4.3-kb nucleotide sequence that separates the initiator codons of the gene pair, is shown by a *stippled rectangle*, and 3'-downstream regions are indicated by *diagonally hatched rectangles*. The direction of transcription (5' → 3') is indicated for each gene. Lengths of sequence components can be estimated from the *scale* below the bar diagram

Truncation of the Globin Gene Complex in Antarctic Icefishes

The fifteen known species of icefishes all share the hemoglobinless condition [2, 21, 22]. As a first step toward systematic evaluation of globin gene status in the channichthyids, we compared the hybridization patterns of notothenioid adult globin cDNAs to the genomes of three icefish species (*Chaenocephalus aceratus, Champsocephalus gunnari*, and *Chionodraco rastrospinosus*) and of four red-blooded relatives [10]. Figure 2 shows that the presence of α1-globin-related DNA sequences (panel A) and the apparent absence of β-globin genes (panel B) are features common to icefish genomes that represent both primitive and advanced genera. By contrast, three red-blooded Antarctic notothenioids [the rockcods *Gobionotothen gibberifrons* and *N. coriiceps* (family Nototheniidae), and the dragonfish *Parachaenichthys charcoti* (family Bathydraconidae)] and a temperate nototheniid (the New Zealand black cod *Notothenia angustata*) gave strong hybridization signals for both α- and β-globin probes. These results suggest that the establishment of the hemoglobinless phenotype preceded the evolutionary radiation of the icefish genera and involved, at a minimum, the deletion of adult β-globin genes.

Its β-globin gene partner being absent, what has become of the adult α1-globin gene of icefishes? One can envision at least three evolutionary scenarios: (1) icefishes continue to express adult α1-globin as an evolutionary relic; (2) the α1-globin gene has been recruited to produce a protein that serves a different function; or (3) the gene has mutated to an inactive state without com-

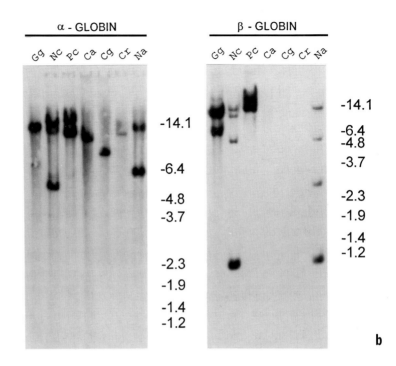

Fig. 2a,b. Globin-related sequences in the genomes of red- and white-blooded antarctic fishes. Southern blots of genomic DNAs from three nototheniids (*Gg*, *G. gibberifrons*; *Na*, *N. angustata*; *Nc*, *N. coriiceps*), a bathydraconid (*Pc*, *P. charcoti*), and three channichthyids (*Ca*, *C. aceratus*; *Cg*, *C. gunnari*; *Cr*, *C. rastrospinosus*) were probed with *N. coriiceps* cDNAs for α globin (**a**) or for β globin (**b**). DNAs were digested with *Bam*HI. The molecular weights of DNA markers are indicated on the vertical axes in kilobase pairs. See Cocca et al. [10] for hybridization and wash conditions. (Reprinted with permission from Cocca et al. [10]. Copyright 1995 by The National Academy of Sciences of the USA)

plete loss from the genome. The first two possibilities were ruled out by assessment of steady-state α1-globin mRNA levels in major tissues and organ systems of *C. aceratus* [10]. α1-Globin transcripts were not detected in any of the icefish tissues and organs, including the cellular component (leukocytes) of blood, the hematopoietic organs (head kidney, trunk kidney, and spleen), heart, gill, liver, white skeletal muscle, and brain. By contrast, α-globin mRNAs were present in abundance in erythrocytes, head kidney, and spleen of the red-blooded *N. coriiceps*. Thus, the most plausible fate of the icefish α1-globin gene is mutation to transcriptional inactivity without extinction of the sequence from the channichthyid genome.

To investigate this last possibility, Zhao et al. [11] isolated and sequenced α1-globin genomic clones from *C. aceratus* and from *C. rastrospinosus*. Figure 3

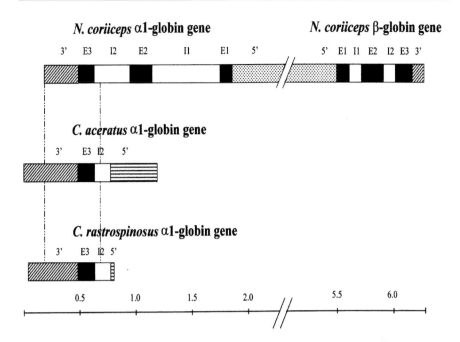

Fig. 3. Comparison of the α1-globin gene remnants of *C. aceratus* and *C. rastrospinosus* to the α1/β-globin gene complex of *N. coriiceps*. The exons (*E1-E3*), introns (*I1, I2*), intergenic region, and 3'-downstream regions of the wild-type *N. coriiceps* globin complex are represented by *filled, open, stippled,* and *hatched rectangles*, respectively. Sequences adjacent to the 5' limits of the icefish α1-globin gene remnants are shown by *horizontally striped boxes*. Box dimensions encompass the extent of sequencing of each gene. (Adapted with permission from Zhao et al. [11])

shows that the icefish α1-globin genes are truncated variants of the *N. coriiceps* α1-globin gene that contain part of intron 2, the entirety of exon 3, and the 3'-untranslated region. The apparent 5' chromosomal breakpoint within intron 2 (between positions 931 and 932) is identical in the two icefish genes, and the 5'-flanking sequences preceding the breakpoint are unrelated to any portion of the *N. coriiceps* α1-globin gene. Beyond the second polyadenylation signal, the icefish and rockcod genes share sequence similarity for at least 180 bp. (Control experiments indicated that the failure to recover the 5'-portion of the α1-globin gene from the icefish genomes did not result from expansion of globin intron 2 in the icefish lineage [11].) These observations are consistent with deletional loss of 5'-upstream α1-globin sequences, including the 5'-untranslated region, exons 1 and 2, intron 1 and part of intron 2, prior to divergence of these two relatively advanced icefish species. Determination of the status of the α1-globin gene in the ancestral channichthyid will require analysis of more primitive icefish species.

Mutational Drift of the α-Globin Gene Remnants of Icefishes: Calibration of the Mutational Clock for Nuclear Genes

Together, the results presented above demonstrate that failure of the icefishes to synthesize the α1 and β globins is due to deletional loss of the 5' end of the notothenioid α1-globin gene and the complete β-globin gene. The residual ice-fish α1-globin gene, no longer under positive selection pressure, subsequently experienced random mutational drift. Figure 4 shows the alignment of the α1-globin gene remnants of *C. aceratus* and *C. rastrospinosus* with the corre-sponding region of the α1-globin gene of *N. coriiceps*. In regions of overlap, the *C. aceratus* and *N. coriiceps* sequences are 96.8% similar, whereas those of *C. rastrospinosus* (the more advanced icefish) and *N. coriiceps* are 95.9% similar. The two icefish sequences, in turn, share 99.0% similarity. Although strongly related to the *N. coriiceps* sequence, the icefish α1-globin remnants share a number of deletions, insertions, and point mutations with respect to the for-mer. For example, the icefish genes share a 16 nucleotide deletion within the 3'-untranslated region (see *N. coriiceps* position 1286) and a five nucleotide inser-tion after *N. coriiceps* position 1441. Furthermore, the sequences of the *C. acer-atus* and *C. rastrospinosus* remnants contain 14 and 17 nucleotide substitu-tions, respectively, relative to the *N. coriiceps* α1-globin sequence. Seven sub-stitutions shared by the icefishes occur within exon 3, but no nonsense codons are introduced into the coding sequence. Transversions exceed transitions in both icefish species (11 transversions and 3 transitions for *C. aceratus*, 14 and 3 for *C. rastrospinosus*). The two icefish remnants have also diverged from each other, as shown by the two nucleotide insertion and three nucleotide substitu-tions in the *C. rastrospinosus* sequence between positions 1010 and 1016 and the single nucleotide deletion after position 1398.

Using the mitochondrial DNA-based age of radiation of the notothenioid fishes, 7-15 million years ago [23], one can estimate the rate of icefish nuclear gene divergence from the frequency of transversion substitutions in the α1-globin gene remnant [12, 24]. *C. aceratus* and the more advanced *C. rastro-spinosus* have accumulated 11 and 14 transversions, respectively, in this frag-ment. Therefore, the estimated rate of icefish nuclear gene divergence in the absence of selective pressure falls in the range 0.12%-0.33% per million years. This slow rate of nuclear gene divergence is consistent with the low specific metabolic rates of the cold-adapted notothenioids [12].

Mechanism of Globin Gene Loss in the Icefishes

The demonstration that some icefishes have undergone deletion of the adult β-globin gene [10] and 5' truncation of the major adult α-globin gene [11] at a single chromosomal site suggests a simple mechanism for loss of the adult glo-bin genes. Figure 5 shows that a single deletional event (scheme X) in the ancestral channichthyid, whose breakpoints are located within intron 2 of the

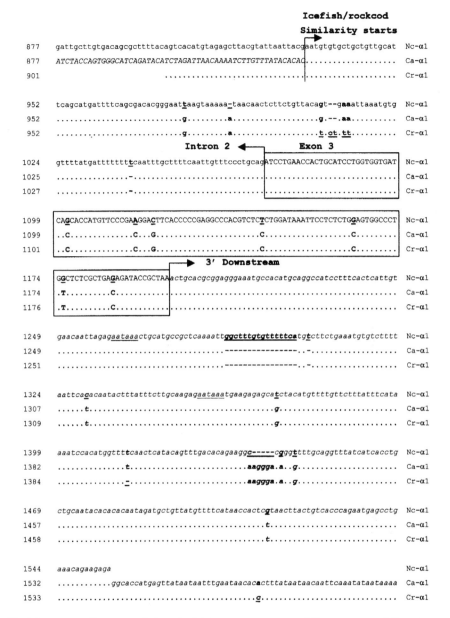

Fig. 4. Sequence alignment of the α1-globin gene remnants of *C. aceratus* and *C. rastro-spinosus* to the α1-globin gene of *N. coriiceps*. The icefish sequences (Ca-α1, Cr-α1) are numbered to correspond to the region of the *N. coriiceps* α1-globin gene (Nc-α1) that begins at nucleotide 877, and the similarity of the three sequences commences at residue 932. Residues identical to the *N. coriiceps* sequence (or to the *C. aceratus* sequence after position 1543) are indicated by *periods*. Nucleotides at positions that differ among the three fishes are shown in *boldface*, and those residues that differ from the consensus are *underlined*. *Dashes* indicate deletions. Exon 3 is enclosed by the *large box* (Copyright 1998 by the American Society for Biochemistry and Molecular Biology, Inc.)

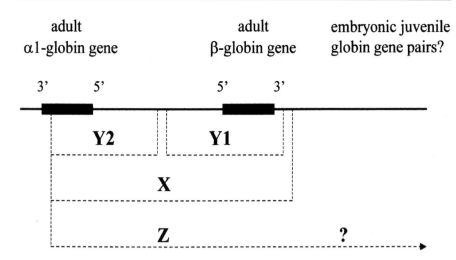

Fig. 5. Potential mechanisms of globin gene deletion by the Antarctic icefishes. Mechanism 1: Simultaneous deletion of the β-globin gene, the intergenic region, and the 5' portion of the linked α1-globin gene (X), followed by random mutation of the α1-globin gene remnant. Mechanism 2: Independent deletion of the β-globin gene (Y1) and of the 5' portion of the linked α1-globin gene (Y2). If embryonic and juvenile α/β-globin gene pairs are linked in the notothenioid genome to the adult globin gene pair, then either model (e.g., single deletion scenario Z) can be extended easily to encompass their probable loss in icefish genomes. (Copyright 1998 by the American Society for Biochemistry and Molecular Biology, Inc.)

α1 globin gene and downstream of the 3'-untranslated region of the β-globin gene, would abolish the expression of the adult globins. Alternatively, multiple deletions (Y1, Y2) that occurred prior to diversification of the icefish clade to yield the disrupted globin locus are a formal, but less likely, possibility. The evolutionary fate of embryonic and juvenile globin genes in icefish genomes is unknown. However, the linkage of embryonic and adult globin genes observed in the zebrafish [8] raises the possibility that the single deletional event postulated here (Fig. 5, scheme Z) might have eliminated coincidentally the embryonic and juvenile genes of the notothenioid globin gene complex.

Genomic Plasticity and the Notothenioid Radiation

The ancestral notothenioid stock probably arose as a sluggish, bottom-dwelling teleost that evolved some 40-60 million years ago in the then temperate shelf waters of the Antarctic continent [2, 25, 26]. With the local extinction of most of the temperate Tertiary fish fauna as the Southern Ocean cooled, the suborder Notothenioidei experienced a dramatic radiation of species, dating from the mid-Miocene (approximately 7-15 million years ago) [2, 23], that enabled it to exploit the diverse habitats provided by a now chronically freezing marine environment. Substantial and rapid reconfiguration of notothenioid genomes, including the evolution of genes encoding antifreeze glycoproteins from a trypsinogen gene precursor [27-29] and the expansion of the tubulin gene families [30], has been pivotal in facilitating both phyletic diversification and cold adaptation within this clade.

The globin and myoglobin gene families of Antarctic icefishes provide further insight into the genetic mechanisms that have facilitated notothenioid speciation. As reviewed here, abolition of the synthesis of hemoglobin by icefishes was likely caused by a single, large deletional event that removed the adult globin gene complex from the genome of the ancestral icefish. By contrast, loss of myoglobin gene expression by icefishes has occurred recurrently by more subtle genomic modifications. In a survey of eight icefish species, Sidell et al. [31] demonstrated that cardiac myoglobin expression has been lost independently on at least three occasions. Furthermore, loss of myoglobin expression has occurred by at least two molecular mechanisms: (1) failure to transcribe the myoglobin gene and/or process the transcripts; and (2) failure to translate myoglobin mRNA to yield the myoglobin polypeptide. The mutational events responsible for loss of myoglobin expression are likely to be small oligonucleotide insertions and/or deletions that disrupt promoter function [32] or produce translational frameshifting within the mRNA [33]; large scale deletion of myoglobin sequences has not occurred [31, 32].

The Antarctic notothenioids are likely to provide many other examples of rapid genomic change, on both large and small scales. The numerous cardiovascular and hematopoietic adaptations of the icefishes, in particular, should provide excellent experimental targets for evaluating the genetic processes that generate diversification in a cold, isolated marine environment.

Acknowledgements

I gratefully acknowledge the many individuals who have contributed to the work that I have synthesized here: my students David Lau, Manoja Ratnayake-Lecamwasam, Amna Saeed, and Yuqiong Zhao; my technician Sandra Parker; and my colleagues Drs. Laura Camardella, Maria Ciaramella, Ennio Cocca, and Guido di Prisco of the Istituto di Biochimica delle Proteine ed Enzimologia, Consiglio Nazionale delle Ricerche, Napoli, Italia. I also acknowledge the logistic support provided to my Antarctic field research program, performed at Palmer Station and in the seas of the Palmer Archipelago, by the staff of the Office of Polar Programs of the National Science Foundation, by the personnel of Antarctic Support Associates, and by the captains and crews of the *R/V Polar Duke* and the *ARSV Laurence M. Gould*. This work was supported by National Science Foundation Grants OPP-9120311, OPP-9420712, and OPP-9815381 to H.W. Detrich.

References

1. Ruud JT (1954) Vertebrates without erythrocytes and blood pigment. Nature 173:848-850
2. Eastman JT (1993) Antarctic fish biology: evolution in a unique environment. Academic, San Diego
3. D'Avino R, Caruso C, Camardella L, Schininà ME, Rutigliano B, Romano M, Carratore V, Barra D, di Prisco G (1991) An overview of the molecular structure and functional properties of the hemoglobins of a cold-adapted Antarctic teleost. In: di Prisco G (ed) Life under extreme conditions: biochemical adaptation. Springer, Berlin Heidelberg New York, pp 15-33
4. di Prisco G, D'Avino R, Caruso C, Tamburini M, Camardella L, Rutigliano B, Carratore V, Romano M (1991) The biochemistry of oxygen transport in red-blooded Antarctic fish. In: di Prisco G, Maresca B, Tota B (eds) Biology of Antarctic fish. Springer, Berlin Heidelberg New York, pp 263-281
5. Macdonald JA, Montgomery JC, Wells RMG (1987) Comparative physiology of Antarctic fishes. Adv Mar Biol 24:321-388
6. di Prisco G (1998) Molecular adaptations of Antarctic fish hemoglobins. In: di Prisco G, Pisano E, Clarke A (eds) Fishes of Antarctica: a biological overview. Springer, Milano
7. McMorrow T, Wagner A, Deryckere F, Gannon F (1996) Structural organization and sequence analysis of the globin locus in Atlantic salmon. DNA Cell Biol 15:407-414
8. Chan FY, Robinson J, Brownlie A, Shivdasani RA, Donovan A, Brugnara C, Kim J, Lau BC, Witkowska E, Zon LI (1997) Characterization of adult α- and β-globin genes in the zebrafish. Blood 89:688-700
9. Miyata M, Aoki T (1997) Head-to-head linkage of carp α- and β-globin genes. Biochim Biophys Acta 1354:127-133
10. Cocca E, Ratnayake-Lecamwasam M, Parker SK, Camardella L, Ciaramella M, di Prisco G, Detrich HW, III (1995) Genomic remnants of α-globin genes in the hemoglobinless antarctic icefishes. Proc Natl Acad Sci USA 92:1817-1821
11. Zhao Y, Ratnayake-Lecamwasam M, Parker SK, Cocca E, Camardella L, di Prisco G, Detrich HW, III (1998) The major adult α-globin gene of Antarctic teleosts and its remnants in the hemoglobinless icefishes: calibration of the mutational clock for nuclear genes. J Biol Chem 273:14745-14752
12. Martin AP, Palumbi SR (1993) Body size, metabolic rate, generation time, and the molecular clock. Proc Natl Acad Sci USA 90:4087-4091
13. Efstratiadis A, Posakony JW, Maniatis T, Lawn RM, O'Connell C, Spritz RA, DeRiel JK, Forget BG, Weissman SM, Slightom JL, Blechl AE, Smithies O, Baralle FE, Shoulders CC, Proudfoot NJ (1980) The structure and evolution of the human β-globin gene family. Cell 21:653-668
14. Lawn RM, Efstratiadis A, O'Connell C, Maniatis T (1980) The nucleotide sequence of the human β-globin gene. Cell 21:647-651
15. Liebhaber SA, Gootsens MJ, Kan YW (1980) Cloning and complete nucleotide sequence of human 5' alpha-globin gene. Proc Natl Acad Sci USA 77:7054-7058
16. D'Avino R, di Prisco G (1989) Hemoglogin from the antarctic fish Notothenia coriiceps neglecta. 1. Purification and characterization. Eur J Biochem 179:699-705
17. Fago A, D'Avino R, di Prisco G (1992) The hemoglobins of Notothenia angustata, a temperate fish belonging to a family largely endemic to the Antarctic Ocean. Eur J Biochem 210:963-970
18. Proudfoot NJ, Maniatis T (1980) The structure of a human α-globin pseudogene and its relationship to α-globin gene duplication. Cell 21:537-544

19. Lacy E, Maniatis T (1980) The nucleotide sequence of a rabbit β-globin pseudogene. Cell 21:545-553

20. Vanin EF (1985) Processed pseudogenes: characteristics and evolution. Annu Rev Gen 19:253-272

21. Hureau JC, Petit D, Fine JM, Marneux M (1977) New cytological, biochemical, and physiological data on the colorless blood of the Channichthyidae (Pisces, Teleosteans, Peciformes). In: Llano GA (ed) Adaptations within antarctic ecosystems. Smithsonian Institution, Washington, DC, pp 459-477

22. Barber DL, Mills Westermann JE, White MG (1981) The blood cells of the Antarctic icefish Chaenocephalus aceratus Lonnberg: light and electron microscopic observations. J Fish Biol 19:11-28

23. Bargelloni L, Ritchie PA, Patarnello T, Battaglia B, Lambert DM, Meyer A (1994) Molecular evolution at subzero temperatures: mitochondrial and nuclear phylogenies of fishes from Antarctica (suborder Notothenioidei), and the evolution of antifreeze glycopeptides. Mol Biol Evol 11:854-863

24. Martin AP, Naylor GJP, Palumbi SR (1992) Rates of mitochondrial DNA evolution in sharks are slow compared with mammals. Nature 357:153-155

25. DeWitt HH (1971) Coastal and deep-water benthic fishes of the Antarctic. In: Bushnell VC (ed) Antarctic map folio series, folio15. American Geographical Society, New York, pp 1-10

26. Eastman JT (1991) Evolution and diversification of Antarctic notothenioid fishes. Am Zool 31:93-109

27. Chen L, DeVries AL, Cheng CHC (1997) Evolution of antifreeze glycoprotein gene from a trypsinogen gene in Antarctic notothenioid fish. Proc Natl Acad Sci USA 94:3811-3816

28. Cheng CHC (1998) Origin and mechanism of evolution of antifreeze glycoproteins in polar fishes. In: di Prisco G, Pisano E, Clarke A (eds) Fishes of Antarctica: a biological overview. Springer, Milano, pp 311-328

29. Cheng CHC, Chen L (1999) Evolution of an antifreeze glycoprotein: a blood protein that keeps Antarctic fish from freezing arose from a digestive enzyme. Nature 401:443-444

30. Parker SK, Detrich HW III (1998) Evolution, organization, and expression of α-tubulin genes in the Antarctic fish Notothenia coriiceps: a tandem gene cluster generated by recent gene duplication, inversion, and divergence. J Biol Chem 273:34358-34369

31. Sidell BD, Vayda ME, Small DJ, Moylan TJ, Londraville RL, Yuan ML, Rodnick KJ, Eppley ZA, Costello L (1997) Variable expression of myoglobin among the hemoglobinless Antarctic icefishes. Proc Natl Acad Sci USA 94:3420-3424

32. Small DJ, Vayda ME, Sidell BD (1998) The myoglobin gene of Antarctic teleosts contains three A+T rich introns. J Mol Evol 47:156-166

33. Vayda ME, Small DJ, Yuan M-L, Costello L, Sidell BD (1997) Conservation of the myoglobin gene among Antarctic notothenioid fishes. Mol Mar Biol Biotech 6:207-216

Oxygen-Transport System and Mode of Life in Antarctic Fish

M. Tamburrini, G. di Prisco

Introduction

The separation of Antarctica from South America occurred 22-25 million years ago with the opening of the Drake Passage, and produced the Circum-Antarctic Current and the development of the Antarctic Polar Front. With the reduction of heat exchange from northern latitudes, cooling of the environment proceeded to the present extreme conditions. To date Antarctica is indeed a unique natural laboratory for the study of temperature adaptations and for understanding the interplay among biochemical/physiological processes, ecology and adaptive evolution.

The physico-chemical features of the Antarctic marine environment have been gradually changing over the past 30-40 million years, in parallel with diversification (and subsequent colonisation) of the teleost suborder Notothenioidei. Extreme temperature conditions were developed and fish became cold adapted. Although stability and constancy characterise the physico-chemical features of the Antarctic marine environment at the present time, the overall process has induced biological diversity among fish, produced by genetic change and evolutionary processes which are reflected at the molecular level.

Haemoglobin (Hb), a direct link between the exterior and the requirements of the body, adapted its functional features in response to major evolutionary pressures. The search for a correlation between fish haematology and the extreme conditions of the Antarctic environment justifies the study of the biochemistry of oxygen transport, giving special attention to biodiversity. This further suggests investigating the relationship between the molecular structure and the oxygen-binding properties of Hb, on the one hand [1-4], and the eco-

Institute of Protein Biochemistry and Enzymology, CNR, Naples, Italy

G. di Prisco, B. Giardina, R.E. Weber (Eds)
Hemoglobin Function in Vertebrates.
Molecular Adaptation in Extreme and Temperate Environments
© Springer-Verlag Italia 2000

logical constraints, on the other. In view of the role of temperature in modifying the oxygenation-deoxygenation cycle in respiring tissues, a thermodynamic analysis is warranted.

Fishes of the perciform suborder Notothenioidei, mostly confined within Antarctic and sub-Antarctic waters, are the dominant component of the Southern Ocean fauna. They comprise 95 of the 174 benthic species of the continental shelf and upper continental slope, and 120 (including non-Antarctic notothenioids) of the 274 Southern Ocean species described to date [5]. Notothenioids probably appeared in the early Tertiary [6] and began to diversify while gradually adapting to progressive cooling and isolation.

Bovichtidae, Pseudaphritidae, Eleginopsidae, Nototheniidae (with the highest number of species, 34 Antarctic and 15 non-Antarctic), Harpagiferidae, Artedidraconidae, Bathydraconidae and Channichthyidae are the families of the suborder. Notothenioids are red-blooded, except Channichthyidae, the only known vertebrates whose blood - in the adult stage - is devoid of Hb [7].

Antarctic Notothenioidei

Having to cope with temperatures below the freezing point of the body fluids, suitable physiological and biochemical specialisations developed in Antarctic fish. In addition to being freezing resistant, they show reduced number of erythrocytes and a reduction in haemoglobin content and multiplicity, This adaptive feature counterbalances the increase in blood viscosity produced by the subzero seawater temperature, with potentially negative physiological effects (e.g. a higher demand for energy needed for circulation). Functional incapacitation of Hb (by means of exposure to carbon monoxide) and reduction of the haematocrit to less than 1% in cannulated fish [8, 9] caused no discernible harm in the absence of metabolic challenges, suggesting that - similar to Hb-less channichthyids - red-blooded fish can carry the routinely required oxygen dissolved in the plasma, and that the need for an oxygen carrier in a stable, cold environment is thus also reduced in red-blooded fish.

Notothenioidei are by far the most thoroughly characterised group of fish in the world. Our studies on their oxygen-transport system have so far been addressed to 38 out of a total of 80 red-blooded Antarctic species. This highly representative number encompasses all major families and also two species of non-Antarctic notothenioids.

Thirty-five species (all bottom dwellers) of the 38 investigated [1, 2] were shown to have a single major form of Hb (Hb 1) and often a minor one (Hb 2, generally having the β chain in common), both displaying - with some exceptions - strong Bohr [10] and Root [11] effects. Each of the remaining three species, all Nototheniidae, have a unique oxygen-transport system, each adjusted to the fish-specific life style, substantially different from that of the sluggish benthic species. *Trematomus newnesi* and *Pagothenia borchgrevinki* are active cryopelagic species; *Pleuragramma antarcticum* is a pelagic, sluggish but migratory fish. Biological diversity is thus reflected at the molecular level of the respiratory process.

The oxygen-transport system of species belonging to the family Artedidraconidae also shows remarkable features which may be the result of life-style adaptation to extreme conditions.

Trematomus newnesi

T. newnesi (family Nototheniidae) actively swims and feeds near the surface [12]. The haematological features differ remarkably from those of the other Antarctic species. The Hb system of *T. newnesi* is made of Hb C, Hb 1 and Hb 2. The amino acid sequence of the globin chains of the three Hbs has been established. Hb 1 has the α chain in common with Hb C, and the β chain in common with Hb 2 [13]. This notothenioid is the only species having two major Hbs (Hb 1 and Hb C), only one of which (Hb C, 20%-25% of the total amount, whereas a mere trace in all other notothenioids) displays pH and organophosphate regulation (Table 1). The Hill coefficient indicates co-operativity of oxygen binding in Hb 1 and Hb 2 in the whole physiological pH range, and loss of co-operativity in Hb C at lower pH (a typical feature of all Root-effect Hbs). Temperature has a slight effect on the oxygen affinity of the most abundant component Hb 1 (the small ΔH of oxygen binding is constant in the pH range 8.0-6.5); thus, no significant amounts of energy are required during the oxygenation-deoxygenation cycle. This energy-saving mechanism may facilitate the function of Hb in the constantly low temperature of the water and

Table 1. The oxygen-transport system of *Trematomus newnesi*

Hb component	% of total	3 mM ATP	Bohr and Root effects	ΔH (kcal/mol oxygen)[a]	
				pH 7.0	pH 8.0
Hb C	20-25	-	Strong	-6	-12
		+	Enhanced	-3	-9
Hb 1	70-75	-	Weak or absent	-4	-4
		+	Not enhanced	-4	-4
Hb 2	3-5	-	Weak or absent	n.d.	n.d.
		+	Not enhanced	n.d.	n.d.

n.d., not determined

[a] The overall oxygenation enthalpy change ΔH (kcal/mol; 1 kcal=4.184 kJ), corrected for the heat of oxygen solubilisation (-3 kcal/mol), was calculated by the integrated van't Hoff equation: $\Delta H = -4.574 \ [(T_1 \cdot T_2)/(T_1 - T_2)] \ \log P_{50}/1000$ (P_{50} is the partial pressure of oxygen required to achieve Hb half saturation).

may reflect molecular adaptation to extreme life conditions [1].

Being a more active fish than the other notothenioids, *T. newnesi* may require functionally distinct Hbs. In fact, this Hb system can ensure oxygen binding at the gills (via Hb 1) and controlled delivery to tissues (via Hb C) even though the active behaviour produces acidosis. High levels of Hb C, conceivably redundant in other notothenioids (which count on Root and Bohr-effect Hb), balance the lack of proton/effector regulation of Hb 1 and Hb 2.

The haemoglobin multiplicity, the oxygen-binding features of Hb 1 and Hb 2, and the presence of functionally distinct components thus reveal that the oxygen transport of *T. newnesi* has unique characteristics.

Pleuragramma antarcticum

This migratory nototheniid is the most abundant and the only fully pelagic species of the high-Antarctic shelf systems. It combines the more general adaptations of all notothenioids [14], with the specialisations necessary to live in the water column. Since the life history and ecology of this species is exceptional among the notothenioids [15], and due to its unique mode of life and great biological significance in the high-Antarctic pelagic system, *P. antarcticum* is highly suitable for studies on adaptation to extreme environmental conditions.

P. antarcticum has three major forms of Hbs (Hb 1, Hb 2 and Hb 3). Among Nototheniidae and the other red-blooded families of Notothenioidei, *P. antarcticum* is the only species with such a high multiplicity of major forms of Hb. Hb 1 has the α chain in common with Hb 2 and the β chain in common with Hb 3. Hb 2 and Hb 3 have no chain in common. The amino acid sequences of the four globins have been established [16, 17]. In Hb 1, a high rate of identity was observed between the chains of *P. antarcticum* and those of other Antarctic species. The chains of Hb 2 and Hb 3 which are not in common with Hb 1 have a high rate of identity with those of other minor Hbs. The identity with globins of non-Antarctic fish Hbs is lower, similar to other Antarctic fish.

In terms of oxygen-binding regulation, the Hbs appeared almost identical. All three Hbs display very strong, effector-enhanced, dependence of oxygen affinity on pH (alkaline Bohr effect). At 2°C the Bohr coefficient ($\phi = \Delta log P_{50}/\Delta pH$) ranges from -0.9 to -1.23. The Root effect is also displayed to the same extent by all three Hbs, and its amplitude is regulated by the physiological effectors organophosphates and chloride, indicating a strong pH dependence of Hb oxygenation in air. Since *P. antarcticum*, as all Antarctic notothenioids, lacks a swim-bladder, Root-effect Hbs may be associated with the occurrence of a choroid rete in the eye which, being poorly vascularised in fish, depends on the diffusion of oxygen from other tissues.

However, the three components display significant differences in thermodynamic behaviour. Hb 1 and Hb 3 show a very strong enthalpy change at pH 8.0 (Table 2), further enhanced by chloride and organophosphates in the for-

Table 2. The oxygen-transport system of *Pleuragramma antarcticum*

Hb component	% of total	3 mM ATP	Bohr and Root effects	ΔH (kcal/ mol oxygen)[a]	
				pH 7.0	pH 8.0
Hb 1	25-30	-	Strong	-12.8	-15.3
		+	Enhanced	-8.6	-17.4
Hb 2	20-25	-	Strong	-3.6	-6.4
		+	Enhanced	-1.8	-8.1
Hb 3	45-50	-	Strong	-0.1	-16.5
		+	Enhanced	-4.1	-7.6

[a] See Table 1

mer, but drastically decreased in the latter; the heat of oxygenation of Hb 2, in the presence and absence of effectors, is much lower. A dramatic decrease is observed at lower pH in Hb 3 and Hb 2 (ΔH approaches zero); in contrast, Hb 1 retains high oxygenation enthalpy, especially when effectors are absent. These observations clearly indicate a stronger Bohr effect at physiological temperatures in Hb 1 (in the presence of the effectors) and Hb 3 (in their absence). In addition, the moderate effect of temperature on Hb 2 in the pH range 7.0-8.0 and on Hb 3 at pH 7.0 suggests that energy-saving mechanisms of oxygen loading and unloading may also become a viable resource for the fish.

The three Hbs of *P. antarcticum*, unlike the minor forms of Hbs found in the benthic notothenioids, cannot be considered as vestigial or larval remnants devoid of physiological significance [1], even though there is a great phylogenetic distance in the sequences between Hb 1 and the globins of Hb 2 and Hb 3 that are not in common. In contrast with other species, in which the expression of multiple genes is observed only in juveniles (di Prisco et al. unpublished), such expression remains high in the adult stage as well, suggesting refined mechanisms of regulation within the gene family.

From a thermodynamic standpoint, the oxygen-transport system of *P. antarcticum* is one of the most specialised such systems ever found in fish. It appears to have been designed to fit its unusual pelagic life style through refined molecular adaptation in the thermodynamic features of each Hb. Although pelagic, this fish is very sluggish. Therefore, rather than having to respond to acidosis, its Hb system responds to the need to optimise oxygen loading and unloading during seasonal migrations through water masses which can have different and fluctuating temperatures. Thus, *P. antarcticum* relies on three major Hbs, which differ functionally mainly in thermodynamic behaviour rather than on pH and organophosphate regulation.

Pagothenia borchgrevinki

The cryopelagic nototheniid *P. borchgrevinki* has a higher Hb concentration than other notothenioids [18]. The high oxygen capacity may correlate with the mode of life of this fish, relatively active in comparison with the other (mostly sluggish bottom-dwellers) notothenioid species. It has five Hbs (Hb C, Hb 0, Hb 1, Hb 2 and Hb 3; Hb 1 is 70-80% of the total, and Hb C only in trace amounts), which were purified and functionally characterised (Table 3) [19]. Intact erythrocytes have a weak Bohr effect. Hb 1, Hb 2 and Hb 3 are functionally similar, with a slightly stronger Bohr effect and a weak Root effect, not significantly influenced by the effectors. Hb 0 has a strong Bohr effect and (disregarding Hb C) is the only component with a strong, effector-enhanced Root effect. The heats of oxygenation of *P. borchgrevinki* Hbs are lower than those of temperate fish Hbs. Differences were detected, in each component, in the absence and presence of the physiological effectors, indicating that temperature variations may have a different effect on the functional properties of each Hb and that chloride and phosphates play an important role in the conformational change between the oxy and deoxy structures.

The complete amino acid sequences of Hb 1 and Hb 0, as well as partial N-terminal or internal sequences of the other Hbs, have been established (Riccio et al., unpublished). Hb 1 and Hb C have the α chain in common with Hb 0 and Hb 2, respectively; Hb 1 has the β chain in common with Hb 2 and Hb 3; and Hb 0 has a different β chain. The α chain of Hb 3 and the β chain of Hb C are currently under investigation.

Table 3. The oxygen-transport system of *Pagothenia borchgrevinki*

Hb component	% of total	3 mM ATP	Bohr and Root effects	ΔH (kcal/ mol oxygen)[a]	
				pH 7.0	pH 8.0
Hb C	≈1	-	Strong	n.d.	n.d.
		+	Enhanced	n.d.	n.d.
Hb 0	5-10	-	Strong	-7.2	-14.7
		+	Enhanced	-5.1	-4.1
Hb 1	70-80	-	Weak	-3.7	-5.5
		+	Not enhanced	-3.7	-4.5
Hb 2	5-10	-	Weak	-3.2	-5.9
		+	Slightly enhanced	-8.0	-7.2
Hb 3	5-10	-	Weak	-5.3	-7.6
		+	Slightly enhanced	n.d.	n.d.

n.d., not determined
[a] See Table 1

The high multiplicity of functionally distinct Hbs indicates that this active, cryopelagic species also has a very specialised Hb system, which may have been developed as an adaptive response to the fish life style and peculiar environmental conditions.

Artedidraconidae

The oxygen-transport system of species of the family Artedidraconidae has been thoroughly investigated. Artedidraconids are benthic fish, have a wide depth distribution, and are largely confined in the Antarctic continental shelf and slope.

The oxygen-binding properties were investigated, in particular with regard to the Hbs of *Artedidraco orianae* and *Pogonophryne scotti* [20]. Both Hbs have an unusually high oxygen affinity and display a relatively small Bohr effect; the Root effect is elicited only by organophosphates and is also reduced. A remarkably unique feature is the absence of co-operative oxygen binding indicated by a Hill coefficient close to one over the whole pH range. These peculiar functional properties are shared almost to the same extent by all the species of the family which have been investigated [2].

Conclusive Remarks

Adaptive strategies in Antarctica are perhaps the central study topic in Antarctic biology. The final stage of the cooling process led to the extreme temperature of the marine environment, which was met by a suite of adaptive features very different from those found in terrestrial organisms.

Some of the fish Hbs display great differences in selected functional characteristics despite a high rate of identity in their primary structure. Isolation and extreme cold make the ensemble of Antarctic fish Hbs a greatly simplified system, which makes it possible to study structure/function relationships not only to address intriguing questions pertaining to the Antarctic extreme environment (e.g. the lack of subunit co-operativity found in many Antarctic Hbs [20]), but also questions of a more general nature (e.g. the molecular basis of the Root effect, an important functional feature of fish Hb).

It has become possible to correlate sequence, multiplicity and oxygen-binding features (thermodynamics, in particular) with ecological constraints, as shown by the Hb systems of *T. newnesi*, *P. borchgrevinki* and *P. antarcticum*, whose life style differs from that of the benthic species. Their blood contains multiple, functionally distinct Hbs, which may be the result of adaptation to the environmental changes experienced by these fish. Correlation with the life style of these species can be deduced [21], since the selective advantage of multiple Hb genes is clear, whereas it is often difficult to establish whether multiple genes are merely surviving, selectively neutral gene duplication.

The very peculiar oxygen-binding properties suggest a less critical role of Hb in the sluggish Artedidraconidae. The physiological role of Hb might merely be that of an "oxygen store" when these fish encounter anoxic conditions. Although the lack of subunit co-operativity resembles that shown by multimeric Hbs of primitive vertebrates (e.g. hagfish, lamprey), the Hbs of Artedidraconidae must be regarded as "modern", in view of their quaternary structure and high sequence identity with major Hbs (which do show co-operative interactions) of other notothenioids [20]. As a consequence, interesting questions arise as to the evolution and mode of function of multi-subunit molecules.

In addition to strictly ecological considerations, such as species number and distribution of individuals among the species [22], the findings summarised in this chapter encourage the study of "physiological biodiversity" as one of the most interesting and useful tools available to gain a better understanding of the biology of Antarctic organisms.

Acknowledgements

This study was conducted as part of the Italian National Programme for Antarctic Research. Thanks are due to V. Carratore, E. Cocca, R. D'Avino, A. Riccio and M. Romano for their contributions, and to the Alfred Wegener Institute, Bremerhaven, Germany, for the participation of the authors in the expedition ANT X/3 (March-May 1992) in the northeastern Weddell Sea.

References

1. di Prisco G, D'Avino R, Caruso C, Tamburrini M, Camardella L, Rutigliano B, Carratore V, Romano M (1991) The biochemistry of oxygen transport in red-blooded Antarctic fish. In: di Prisco G, Maresca B, Tota B (eds) Biology of Antarctic fish. Springer, Berlin Heidelberg New York, pp 263-281
2. di Prisco G, Tamburrini M, D'Avino R (1998) Oxygen-transport systems in extreme environments: multiplicity and structure/function relationship in hemoglobins of Antarctic fish. In: Pörtner HO, Playle R (eds) Cold ocean physiology. Cambridge University Press, Cambridge, pp 143-165 (Society of experimental biology, seminar series 66)
3. di Prisco G, Giardina B (1996) Temperature adaptation: molecular aspects. In: Johnston IA, Bennett AF (eds) Animals and temperature. Phenotypic and evolutionary adaptation. Cambridge University Press, Cambridge, pp 23-51 (Society of experimental biology, seminar series 59)
4. di Prisco G (1997) Physiological and biochemical adaptations in fish to a cold marine environment. In: Battaglia B, Valencia J, Walton DWH (eds) Proc SCAR 6th Biol Symp, Venice (Antarctic communities: species, structure and survival). Cambridge University Press, Cambridge, pp 251-260
5. Gon O, Heemstra PC (eds) (1990) Fishes of the Southern Ocean. JLB Smith Institute of Ichthyology, South Africa
6. Eastman JT (1993) Antarctic fish biology. Evolution in a unique environment. Academic, San Diego

7. Ruud JT (1954) Vertebrates without erythrocytes and blood pigment. Nature 173:848-850
8. Wells RMG, Macdonald JA, di Prisco G (1990) Thin-blooded Antarctic fishes: a rheological comparison of the haemoglobin-free icefishes *Chionodraco kathleenae* and *Cryodraco antarcticus* with a red-blooded nototheniid, *Pagothenia bernacchii*. J Fish Biol 36:595-609
9. di Prisco G, Macdonald JA, Brunori M (1992) Antarctic fish survive exposure to carbon monoxide. Experientia 48:473-475
10. Riggs AF (1988) The Bohr effect. Ann Rev Physiol 50:181-204
11. Brittain T (1987) The Root effect. Comp Biochem Physiol 86B:473-481
12. Eastman JT (1988) Ocular morphology in Antarctic notothenioid fishes. J Morphol 196:283-306
13. D'Avino R, Caruso C, Tamburrini M, Romano M, Rutigliano B, Polverino de Laureto P, Camardella L, Carratore V, di Prisco G (1994) Molecular characterization of the functionally distinct hemoglobins of the Antarctic fish *Trematomus newnesi*. J Biol Chem 269:9675-9681
14. Tamburrini M, di Prisco G (1993) Biochemical adaptations in polar marine environments. Ital J Biochem 42:258-259A
15. Hubold G (1985) On the early life history of the high-Antarctic silverfish *Pleuragramma antarcticum*. In: Siegfried WR, Condy PR, Laws RM (eds) Proc 4th SCAR Biol Symp (Antarctic nutrient cycles and food webs). Springer, Berlin Heidelberg New York, pp 445-451
16. Tamburrini M, D'Avino R, Fago A, Carratore V, Kunzmann A, di Prisco G (1996) The unique hemoglobin system of *Pleuragramma antarcticum*, an Antarctic migratory teleost. Structure and function of the three components. J Biol Chem 271:23780-23785
17. Tamburrini M, D'Avino R, Carratore V, Kunzmann A, di Prisco G (1997) The hemoglobin system of *Pleuragramma antarcticum*: correlation of hematological and biochemical adaptations with life style. Comp Biochem Physiol 118A:1037-1044
18. Macdonald JA, Wells RMG (1991) Viscosity of body fluids from Antarctic notothenioid fish. In: di Prisco G, Maresca B, Tota B (eds) Biology of Antarctic fish. Springer, Berlin Heidelberg New York, pp 163-178
19. Tamburrini M, di Prisco G (1994) The unique features of the hemoglobin system of the Antarctic teleost *Pagothenia borchgrevinki*. "Proteine '94", Abstr A25, p 43
20. Tamburrini M, Romano M, Carratore V, Kunzmann A, Coletta M, di Prisco G (1998) The hemoglobins of the Antarctic fishes *Artedidraco orianae* and *Pogonophryne scotti*. Amino acid sequence, lack of cooperativity, and ligand binding properties. J Biol Chem 273:32452-32458
21. di Prisco G, Tamburrini M (1992) The hemoglobins of marine and freshwater fish: the search for correlations with physiological adaptation. Comp Biochem Physiol 102B:661-671
22. Arntz WE, Gutt J, Klages M (1997) Antarctic marine biodiversity. In: Battaglia B, Valencia J, Walton DWH (eds) Proc SCAR 6th Biol Symp, Venice (Antarctic communities: species, structure and survival). Cambridge University Press, Cambridge, pp 3-14

Functional Properties of the Cathodic Hemoglobin Component from Two Species of Anguilliformes

A. Olianas[1], M.T. Sanna[1], A. Fais[1], A. Pisano[1], S. Salvadori[2], A.M. Deiana[2], M. Corda[1], M. Pellegrini[1]

Introduction

Comparative studies of haemoglobin (Hb) function are of special interest since they may lead to an understanding of those changes which, in the course of evolution, have developed in different organisms to meet specific physiological requirements.

Unlike the majority of mammals, which only produce a single major Hb component (>90% of the Hb content of the red cell), many fish species have multiple Hb components which show considerable differences in amino acid residue sequence and functional properties [1, 2]. Despite the considerable effort which has been devoted to characterizing a number of fish Hbs, no unifying theory has been proposed concerning the biological significance of this multiplicity. In fact, these Hb components may or may not show functional differences. For instance, the two major Hbs from carp are not functionally distinguishable from each other [3], while trout and eel Hb systems are characterized by two types of Hb which differ markedly in their structural and functional properties [4-6]; these latter multiple Hbs have been called cathodic and anodic Hbs on the basis of their electrophoretic behaviour. While the electrophoretically anodic Hb components with low isoelectric points that are encountered in all species have relatively low O_2 affinities and pronounced pH sensitivities (Bohr and Root effects), the cathodic Hbs found in some species exhibit high affinities and low pH dependence. Clearly, functional heterogeneity of several Hb types within the red cells might extend the range of conditions under which oxygen can be transported effectively around the blood stream and may allow a division of labour between the various components so that each fulfils a specific role [7].

[1]Department of Science Applied to Biosystems and [2]Department of Animal Biology and Ecology, University of Cagliari, Cagliari, Italy

G. di Prisco, B. Giardina, R.E. Weber (Eds)
Hemoglobin Function in Vertebrates.
Molecular Adaptation in Extreme and Temperate Environments
© Springer-Verlag Italia 2000

A number of crystal structures of R- and T-state fish Hbs have been described [8, 9]: analysis of these results have demonstrated that the basic features of the mechanism proposed by Perutz and co-workers for allosteric transition in human Hb also apply to fish Hbs, despite the considerable difference in amino acid sequence [10]. In many cases, the functional effects of several heterotropic ligands and/or the mutation of selected residues have also been rationalized in terms of their structural effects. However, a rationale to account for the subtle action of pH is still lacking [11].

In this chapter we report on the functional characterization of the cathodic Hb component from two species of moray that live in the Mediterranean Sea, the brown moray *Gymnothorax unicolor* and the moray eel *Muraena helena,* which are both Anguilliformes. Our aim is to contribute to the understanding of the biological significance of Hb multiplicity in fish and of structure-function relationships in gas-binding proteins in an attempt to correlate the findings at the molecular level with physiological adaptations.

Materials and Methods

G. unicolor and *M. helena* specimens were caught off the coast of Sardinia. Blood samples were collected by venepuncture from specimens anaesthetized by tricaine methane sulphonate (MS222) using heparinized syringes. Immediately after sampling, erythrocytes were pelleted by centrifugation for 5 min at 1000 g, then washed twice through cycles of resuspension and centrifugation at 1000 g, using cold isotonic NaCl solution. Lysis was obtained by adding 2 volumes of cold hypotonic buffer (1.0 mM Tris/HCl pH 8.0). Stroma were removed by centrifugation at 12,000 g for 30 min. Isoelectric focusing (IEF) experiments on thin-layer 5% polyacrylamide slab gels (pH range 3.5-10.0) were performed according to Manca et al. [12] and Masala and Manca [13]. Gels were stained for proteins using a 0.1% (mass/vol.) bromophenol blue solution.

Removal of organic phosphate (i.e. stripped Hb) was obtained by passing haemolysates through a Sephadex G-25 column (25–2.5 cm) equilibrated with 10 mM Hepes buffer pH 8.0 containing 0.1 M NaCl. Hb components were isolated by ion-exchange chromatography on a column (2.5–20 cm) of DEAE-cellulose in 10 mM Tris/HCl pH 7.7. The first component (cathodic Hb) passed unretarded through the column; the anodic components were eluted both by decreasing the pH of the buffer to 7.0 and increasing its concentration to 100 mM.

Dissociated globin chains were separated in polyacrylamide gels in the presence of acetic acid, 8 M urea and Triton X-100 (AUT-PAGE) [14-16].

The globin chains were isolated from the purified Hbs, after incubation in 0.1% trifluoroacetic acid containing 5% 2-mercaptoethanol for 10 min at room temperature, by reverse-phase high performance liquid chromatography (HPLC) on a Nucleosil C_{18} column (Sigma-Aldrich, 0.46–25 cm), equilibrated with eluent A (45% acetonitrile containing 0.3% trifluoroacetic acid), and eluted with a linear 0-100% gradient of eluent B (60% acetonitrile) within 45 min.

Oxygen dissociation curves were obtained spectrophotometrically by the tonometric method [17], using a Varian spectrophotometer model 2300 at a protein concentration of 3-5 mg/ml.

Results

The hemolysates from *M. helena* analysed by IEF display three different patterns, owing to the presence of three phenotypes in the population [18]. All of them contain the same cathodic Hb (Hb I), with isoelectric pH=8.2 [19], while either of the anodic Hbs (Hb II and Hb III), with isoelectric pH=5.92 and 6.34, respectively [19], or both of them may be present in the same phenotype. Phenotypes b, used in this study, only have Hb I and Hb II, while the erythrocyte lysates from *G. unicolor* always show only two Hb components: the cathodic Hb has lower isoelectric point than Hb I from *M. helena*, while the anodic Hb displays a similar electrophoretic pattern to that of Hb III from *M. helena* (unpublished data).

The cathodic Hb fractions of the two species (Fig. 1) were separated from the anodic ones in anion-exchange chromatography, where they eluted unretarded. The elution profiles of the reverse-phase HPLC of the globin chains obtained from the purified cathodic Hbs showed that the α and β chains from *M. helena* are faster than the corresponding chains from *G. unicolor* (Fig. 2). This difference between the four globins is born out by their electrophoretic patterns in AUT-PAGE reported in Figure 3.

The structural differences revealed by chromatographic and electrophoretic analysis also resulted in different functional behaviour. Although the oxygen binding properties of the cathodic component from *M. helena* had already been tested in Bistris or Tris buffer in the presence of 0.1 M NaCl [18], we re-examined the effect of pH on its oxygen affinity and cooperativity in 0.1M Mes/Hepes/Taps NaOH buffer within the 6.5-8.0 pH range and compared it to that of the cathodic Hb from *G. unicolor*. In fact it was noted that this kind of buffer does not influence oxygenation via perturbation of the free chloride activity, as may occur with Bis (Tris) buffers [20], and it also allows O_2 equilibria of Hb solutions to be recorded both in the complete absence of chloride and in the presence of a constant concentration of this anion within a wide pH range. Therefore it was possible to measure the oxygen binding parameters in the absence and in the presence of 100 mM NaCl and/or 1 mM ATP or GTP, which are the physiological modulators of fish Hbs, at 20° C (Fig. 4).

In spite of similar intrinsically high O_2 affinity and low cooperativity in the absence of any effector, the cathodic Hbs of the two species differed with respect to the effect of pH: stripped *G. unicolor* Hb showed a slight reverse Bohr effect, which was enhanced by chloride ions but was obliterated by organic phosphates, while stripped *M. helena* Hb was not affected in its O_2 affinity by

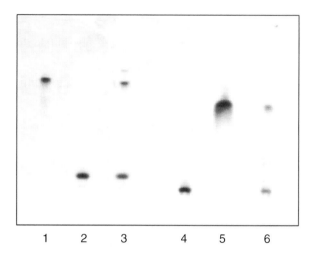

Fig. 1. Isoelectric focusing (IEF) of *G. unicolor* (*lane 3*) and *M. helena* (*lane 6*) haemolysates and their cathodic (*lanes 2 and 4*) and anodic (*lanes 1 and 5*) components separated by DEAE-cellulose chromatography. Thin layer 5% polyacrylamide slab gel electrophoresis was performed according to Manca et al. [8] and Masala and Manca [9]. The cathode was at the bottom and migration was toward the anode; the ampholyte pH range was 3.5-10.0. The gel was stained for proteins using a 0.1% (mass/vol.) bromophenol blue solution

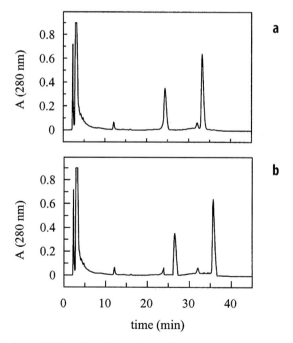

Fig. 2a,b. Reverse-phase HPLC of the globin chains from *G. unicolor* (a) and *M. helena* (b) cathodic component. Details are given in "Materials and Methods"

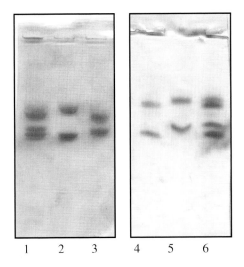

Fig. 3. Acid Urea Triton X-100 polyacrylamide gel electrophoresis (AUT-PAGE) of disso-ciated globin chains from *G. unicolor* (*lane 1*) and *M. helena* (*lane 6*) haemolysates and their cathodic (*lanes 2 and 5*) and anodic (*lanes 3 and 4*) components separated by DEAE-cellulose chromatography. Separation was obtained by PAGE in the presence of acetic acid, 8 *M* urea and Triton X-100 [14-16]. The gel was stained with 0.5% Coomassie brillant blue

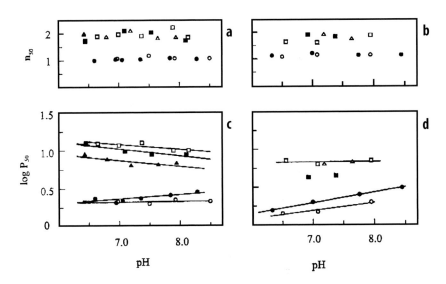

Fig. 4a-d. Effect of pH at 20°C on the oxygen affinity ($\log p_{50}$) and cooperativity (n_{50}) of *M. helena* (**a, c**) and *G. unicolor* (**b, d**) cathodic Hb. Experimental conditions: 0.1 *M* Mes/Hepes/Taps buffer (○), plus 0.1 *M* NaCl (●), in the presence of 1 m*M* GTP (□), plus 0.1 M NaCl (■) or 1 m*M* ATP (△), plus 0.1 M NaCl (▲)

pH changes, even though a small reverse Bohr effect may be elicited by chloride at alkaline pH and a slight normal Bohr effect was observed in the presence of organic phosphates. Upon addition of saturating amounts of GTP, its modulator effect on the O_2 affinity (expressed in terms of $\Delta\log p_{50}$) of *G. unicolor* Hb was about 20% less than that observed on *M. helena* Hb and about 35% less if 100 mM chloride was also added to the solution. Furthermore, both ATP and GTP raised co-operativity, though to differing extents: n_{50} values were 1.7±0.2 and 2.0±0.2 in the case of *G. unicolor* and *M. helena* Hb, respectively. Another important difference between the functional properties of these Hbs was that both ATP and GTP decreased the O_2 affinity of *G. unicolor* Hb to the same extent, whereas GTP was a stronger modulator than ATP of *M. helena* Hb O_2 affinity. However, it should be noted that the presence of both chloride and GTP caused a smaller decrease in O_2 affinity than that induced by GTP or ATP alone.

Discussion

The functional properties of both cathodic Hbs reported here closely match those of other fish which have two functionally different types of Hb [4, 21, 22], called anodic and cathodic Hbs with respect to their isoelectric point. The anodic components are characterized by the Root effect, an unusually strong dependence of oxygen affinity on pH which stabilizes the molecule below pH 6.0 in the low-affinity T-state so strongly that oxygen binding becomes non-co-operative. This allows discharge of oxygen into the swim-bladder against the concentration gradient by the secretion of lactic acid into the blood in the gas gland. The cathodic fractions, in contrast, show very little or no change in oxygen affinity and co-operativity with pH. Therefore, this second type of Hb will predictably secure O_2 loading under internal acidosis induced by exercise stress when the pH at the gills drops to a level that is too low for efficient uptake by the Root-effect anodic Hb [23]. Thus, cathodic Hbs seem to constitute a circulating O_2 reservoir that ensures a sufficient O_2 release at all tissues during concentrated physical activity.

Although the physiological significance of cathodic Hbs appears to be clear, the molecular mechanisms governing their functional behaviour may only be discussed on the basis of the primary structure of two closely related fish Hbs, the β chain of *M .helena* and both the α and β chains of *Anguilla anguilla* cathodic Hb, which are already known [18, 24].

The Bohr effect of human Hb has mainly been ascribed to histidine residue 146β(HC3) which forms a salt-bridge with Asp94β(FG1) in the T-state. This bond is broken as the protein moves to the R-state, lowering the pK$_a$ of the imidazole group and releasing protons under physiological conditions. No such interactions can take place in *M. helena* and *A. anguilla* cathodic Hb, where His146β is replaced by Phe even though Asp94β is still present in the former and substituted by Glu in the latter. The other known Bohr group in human Hb is the N-terminal amino group of the α-chain, which is acetylated in both Hbs and, hence, unable to bind protons in either the T- or R-state. Therefore, as regards the effect of pH, the substitution of His at the C-terminus of the β chain by Phe and the acetylation of the N-terminus of the α-chain brings about inhibition of the alkaline Bohr effect.

Among Anguilliformes, the cathodic Hb component from the Mediterranean moray *M. helena* is insensitive to pH changes [18] while the cathodic Hb from the American and European eels, *A. rostrata* and *A. anguilla,* show a reverse Bohr effect [25, 24]. This effect, if present, indicates proton binding to the Hb upon oxygenation by an increase in the pK value in the transition from the T- to the R-state of some amino acid residues: in the case of human deoxy-HbA, Hisβ143 (H21) has been considered largely responsible for the reverse Bohr effect [26] because the nearby positive charges of Lysβ82(EF6) and Lysβ144(HC1) could lower the pK value of the deoxy-Hb Hisβ143 with respect to that of the oxy-Hb. But if the same effect is still operative at high pH, this could be due, at least in part, to an uncommonly high pK of the 143b residue [27]: the replacement His→Lys in eel and His→Arg in *M. helena* cathodic Hb at this position correlates with this hypothesis. For eels it has also been suggested that the reverse Bohr effect resulting from destabilization of the T-state, which may occur at low pH, could be due to repulsion between the protonated N-termini of the β-chains, of which one may be uncharged at high pH [27]: the rise in the pK of this group correlates with the substitution of the close, positively charged β2-His (in human Hb) by negatively charged glutamate in most fish Hbs. Nevertheless, it should be noted that since the reverse Bohr effect is annihilated by the organic phosphates ATP and GTP in *M. helena* and *G. unicolor* Hb, it cannot be operative at physiological conditions.

The high O_2 affinity displayed by these cathodic Hbs may be due, at least in part, to an altered intrinsic affinity in the α or β subunits due to the substitution of some of the residues in contact with the heme groups. In *M. helena* cathodic Hb the replacement Ser→Gln in position CD3, in comparison with human HbA, could be responsible for an additional hydrogen bond with one of the haeme propionic groups that may alter the subunit affinity, as has been also suggested for the Ser→Lys substitution in eel Hb [24].

The phosphate-binding site in fish Hb includes the N-termini Val(NA1), Glu(NA2), Lys(EF6) and Arg(H21) of the β-chains [28, 29]. Since all these residues are conserved in the cathodic Hb from *A. anguilla* and *M. helena*, this could explain the strong effect on oxygen affinity displayed by GTP and may suggest that these residues contribute to the reverse Bohr effect in the absence of alkaline Bohr groups.

Moreover, the different responses to ATP and GTP modulation could be explained by the presence in both Hbs of Glu in the phosphate-binding pocket at position NA2β, which allows GTP to establish an additional hydrogen bond with respect to ATP, as has been suggested for carp Hb [28, 29]. This hypothesis is confirmed by trout IV, which, in place of Glu, has Asp, whose shorter side chain prevents additional interaction, thereby rendering ATP and GTP identical with regards to protein moiety. In contrast, *M. helena* anodic Hb showed similar responses to ATP and GTP modulation despite the presence of Glu in

NA2β position [18]: since *G. unicolor* cathodic Hb cannot distinguish between ATP and GTP either, knowledge of the amino acid residue present at this site could be vital for a better insight into the molecular basis of this heterotropic modulation in fish Hb.

Interestingly, the presence of both chloride and GTP causes a lower decrease in O_2 affinity than that induced by GTP alone. This indicates that the binding sites for the two anions overlap, at least in part, and suggests the absence of additional sites for allosteric chloride binding in these Hbs as well as in the closely related *A. anguilla* cathodic Hb [24].

Determination of the primary structure of *G. unicolor* cathodic Hb, which is still in progress, should provide new information to understand the structural basis of the special functional properties of Anguilliformes Hbs.

References

1. di Prisco G, Tamburrini M (1992) The hemoglobins of marine and freshwater fish: the search for correlations with physiological adaptation. Comp Biochem Physiol 102B:661-671
2. Weber RE (1982) Intraspecific adaptation of hemoglobin function in fish to environmental oxygen availability. In: Addink ADF, Spronk N (eds) Exogenus and esogenus influences on metabolic and neural control, vol 1. Pergamon, Oxford, pp 87-102
3. Tan AL, De Young A, Noble RW (1972) The pH dependence of the affinity, kinetics, and cooperativity of ligand binding to carp hemoglobin, *Cyprinus carpio*. J Biol Chem 247:2493-2498
4. Binotti I, Giovenco S, Giardina B, Antonini E, Brunori M, Wyman J (1971) Studies on the functional properties of fish hemoglobins. The oxygen equilibrium of the isolated hemoglobin components from trout blood. Arch Biochem Biophys 142:274-280.
5. Brunori M, Bonaventura J, Bonaventura C, Giardina B, Bossa F, Antonini E (1973) Hemoglobins from trout: structural and functional properties. Mol Cell Biochem 1:189-196
6. Weber RE, Lykkeboe G, Johansen K (1976) Physiological properties of eel hemoglobin: hypoxic acclimation, phosphate effects and multiplicity. J Exp Biol 64:75-88
7. Weber RE (1990) Functional significance and structural basis of multiple hemoglobins with special reference to ectothermic vertebrates. In: Truchot JP, Lahlou B (eds) Animal nutrition and transport processes. 2. Transport, respiration and excretion: comparative and environmental aspects. Karger, Basel, pp 58-75 (Comparative physiology, vol 6)
8. Ito N, Komiyama NH, Fermi G (1995) Structure of deoxyhaemoglobin of Antarctic fish *Pagothenia bernacchi* with an analysis of the structural basis of the Root effect by comparison of the liganded and unliganded haemoglobin structures. J Mol Biol 250:648-658
9. Tame JRH, Wilson JC, Weber RE (1996) The crystal structure of trout Hb I in the deoxy and carbonmonoxy forms. J Mol Biol 259:749-760
10. Perutz MF, Fermi G, Luisi B, Shaanan B, Liddington R (1987) Stereochemistry of cooperativity mechanism in haemoglobin. Acc Chem Res 20:309-321

11. Mazzarella l, D'Avino R, di Prisco G, Savino C, Vitagliano L, Moody PCE, Zagari A (1999) Crystal structure of *Trematomus newnesi* haemoglobin re-opens the Root effect question. J Mol Biol 287:897-906

12. Manca L, De Muro P, Masala B (1988) Hb G-Philadelphia, or [α68(E17)Asn$\rightarrow$Lys], in North Sardinia: detection by isoeletric focusing and HPLC of tryptic peptides. Clin Chim Acta 177:231-238

13. Masala B, Manca L (1991) Detection of the common Hb F Sardinia [$^A\gamma$(E19)Ile$\rightarrow$Tyr] variant by isoelectric focusing in normal newborns and in adults affected by elevated fetal hemoglobin syndromes. Clin Chim Acta 198:195-202

14. Efremov GD, Markovska B, Stojanovski N, Petkov G, Nikolov N, Huisman THJ (1981) The use of globin chain electrophoresis in polyacrylamide gels for the quantitation of $^G\gamma$ and $^A\gamma$ ratio in fetal hemoglobin. Hemoglobin 5: 637-651

15. Manca L, Formato M, Demuro P, Gallisai D, Orzalesi M, Masala B (1986) The γ, globin chain heterogeneity of the Sardinian newborn baby. Hemoglobin 10:519-528

16. Braend M, Nesse LL, Efremov GD (1987) Expression and genetics of caprine haemoglobins. Animal Genetics. 18:223-231

17. Giardina B, Amiconi G (1981) Measurament of binding of gaseous and non-gaseous ligands to hemoglobin by conventional spectrophotometric procedures. Methods Enzymol 76:417-427

18. Pellegrini M, Giardina B, Olianas A, Sanna MT, Deiana AM, Salvadori S, di Prisco G, Tamburrini M, Corda M (1995) Structure/function relationships in the hemoglobin components from moray (*Muraena helena*). Eur J Biochem 234:431-436

19. Rizzotti M, Pagni S, Bentivegna F (1990) Conservation of peculiar structural properties by the hemoglobins of anguilloid eels (Teleostei). Z Zool Syst Evolut Forsch 28:12-19

20. Weber RE (1992) Use of ionic and zwitterionic (Tris/BisTris and HEPES) buffers in studies on hemoglobin function. J Appl Physiol 72:1611-1615

21. Hashimoto K, Yamaguchi Y, Matsuura F, (1960) Comparative studies on two hemoglobins of salmon. IV. Oxygen dissociation curve. Bull Jpn Soc Scient Fish 26:827-834

22. Weber RE, Lykkeboe G, Johansen K, (1975) Biochemical aspects of the adaptation of hemoglobin-oxygen affinity of eels to hypoxia. Life Sci 17:1345-1350

23. Brunori M (1975) Molecular adaptations to physiological requirements: the hemoglobin system of trout. In: Horeckor BL, Stadtman ER (eds) Current topics in cellular regulation vol 9, Academic, New York, pp 1-39

24. Fago A, Carratore V, di Prisco G, Feuerlein RJ, Sottrup-Jensen L, Weber RE (1995) The cathodic hemoglobin of *Anguilla anguilla*. J Biol Chem 270:18897-18902

25. Gillen RG, Riggs A (1973) Structure and function of the isolated hemoglobins of the American eel, *Anguilla rostrata*. J Biol Chem 248: 1961-1969

26. Perutz MF, Kilmartin JV, Nishikura K, Fogg JH, Butler PJG (1980) Identification of residues contributing to the Bohr effect of human hemoglobin. J Mol Biol 138:649-670

27. Feuerlein RJ, Weber RE (1996) Oxygen equilibria of cathodic eel hemoglobin analysed in terms of the MWC model and Adair's successive oxygenation theory. J Comp Physiol B 165:597-606

28. Perutz MF, Brunori M (1982) Stereochemistry of cooperative effects in fish and amphibian hemoglobins. Nature 299:421-426

29. Gronenborn AM, Clore GM, Brunori M, Giardina B, Falcioni G, Perutz MF (1984) Stereochemistry of ATP and GTP bound to fish haemoglobins. J Mol Biol 178:731-742

Oxygen Transport in Diving Vertebrates

M. Corda[1], M. Tamburrini[2], M. Pellegrini[1], A. Olianas[1], A. Fais[1],
M.C. De Rosa[3], G. di Prisco[2], B. Giardina[3]

Introduction

Oxygen transport proteins have developed, during evolution, complex regulatory molecular mechanisms to optimize the oxygenation-deoxygenation cycle according to the physiological needs of a given species. Considering the variety of species that depend on hemoglobin for oxygen transport, these molecules must execute their primary function under extreme environmental conditions. In general, these mechanisms appear to be based on a thermodynamic linkage between binding of allosteric effectors (H^+, CO_2 and Cl^-, as well as organic and inorganic phosphates) and the basic reaction of hemoglobin with O_2 [1]. For example, hemoglobins from Arctic mammals are characterized by a very low temperature sensitivity of oxygen binding, which has been interpreted as being beneficial to animals living in cold environments. Thus, with external temperature as low as -40°C, it is vital to reduce the overall ΔH of oxygen binding (generally exothermic, $\Delta H < 0$) as much as possible: deoxygenation will require much less heat and oxygen can still be released from the blood to the colder peripheral tissues. For this reason the hemoglobin of diving animals could be an interesting molecule for study since these mammals are specialized for prolonged dives, often in cold environments. We compared the hemoglobin systems from two species of whale: *Balaenoptera physalus* and *Balaenoptera acutorostrata,* which live in the Mediterranean and the Arctic seas, respectively. These species may have developed specific mechanisms for the maintenance of adequate oxygen supply to cold peripheral tissues in hypoxic conditions and in relation to their habitat.

[1]Dept. of Sciences Applied to Biosystems, University of Cagliari, Monserrato (CA), Italy
[2]Institute of Protein Biochemistry and Enzymology, CNR, Naples, Italy
[3]Institute of Chemistry and Clinical Chemistry and CNR Center for Receptor Chemistry, Catholic University of Rome, Rome, Italy

G. di Prisco, B. Giardina, R.E. Weber (Eds)
Hemoglobin Function in Vertebrates.
Molecular Adaptation in Extreme and Temperate Environments
© Springer-Verlag Italia 2000

Materials and Methods

Blood samples were collected using heparin as anticoagulant. Hemolysates were prepared by lysis of washed erythrocytes with 4 volumes of water and 0.5 volume of toluene, followed by centrifugation. Removal of organic phosphates from purified hemoglobin was obtained using a Sephadex G-25 column equilibrated with 0.01 M Tris HCl pH 8.0 containing 0.1 M NaCl. Oxygen equilibrium curves were determined at 20°C and 37°C, in the pH range 7.0-8.0, using a modified diffusion chamber technique [2, 3]. Serially connected Wösthoff gas mixing pumps generated the stepwise increases in oxygen tension, while absorbance changes between zero and full saturation were monitored at 436 nm with an Eppendorf spectrophotometer model 1101M. The experiments were performed in 100 mM Tris-HCl, containing 100 mM NaCl, in the absence and in the presence of 1% CO_2. The overall oxygenation enthalpy change ΔH (Kcal/mol), corrected for the heat of oxygen solubilization (-3 Kcal/mol), was calculated by the integrated van't Hoff equation:

$$\Delta H = -4.574 \left[(T_1 T_2) / (T_1 - T_2) \right] \Delta \log p_{50} / 1000$$

Results

The pH dependence of the oxygen affinity obtained at 20°C and 37°C of hemoglobin from *Balaenoptera physalus* is shown in Fig. 1 in comparison with that of hemoglobin from *Balaenoptera acutorostrata*. Both whale hemoglobins display a higher affinity for oxygen with respect to human HbA. This is in agreement with the observation that hemoglobins from diving mammals are generally characterized by a higher affinity for oxygen than those from other mammals [4]. The CO_2 effect on the oxygen affinity of *B. physalus* and *B. acutorostrata* hemoglobin, at 20°C and 37°C, is shown in Figs. 2 and 3. At 20°C the CO_2 effect follows a trend which is very similar to that for human HbA [5], showing a substantial increase in oxygen release in the presence of carbon dioxide. The effect is particularly evident under stripped conditions, but it is still clearly observable in the presence of 2,3-Diphosphoglycerate (DPG) 3 mM (data not shown). Moreover, and as expected, the specific effect of carbon dioxide increases at high pH values, where the amino groups involved (pK=7.2) are mainly uncharged. Completely different results were obtained at 37°C. At 37°C, the effect of carbon dioxide on the oxygen affinity of whale hemoglobins disappears both in the absence and in the presence of 2,3-DPG over the whole pH range examined. It is worth recalling that under the same experimental conditions, human hemoglobin is affected by carbon dioxide [3], while in the case of the mole (*Talpa europaea*), a mammal adapted to a hypoxic environment, carbon dioxide has almost no effect on oxygen unloading [6].

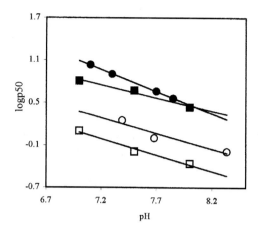

Fig. 1. Oxygen Bohr effect of *Balaenoptera physalus* (*squares*) and *Balaenoptera acutorostrata* (*circles*) hemoglobin, *empty symbols* at 20°C and *filled symbols* at 37°C. Experimental conditions: 0.1 M Tris-HCl, 0.1 M NaCl

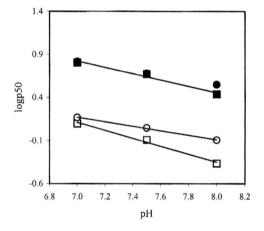

Fig. 2. Oxygen Bohr effect of *Balaenoptera physalus* hemoglobin, (*squares*) stripped, (*circles*) +1% CO_2, *empty symbols* at 20°C and *filled symbols* at 37°C. Experimental conditions: 0.1 M Tris-HCl, 0.1 M NaCl

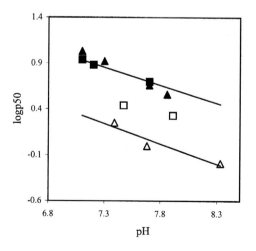

Fig. 3. Oxygen Bohr effect of *Balaenoptera acutorostrata* hemoglobin, (*triangles*) stripped, (*squares*) +1% CO_2, *empty symbols* at 20°C and *filled symbols* at 37°C. Experimental conditions: 0.1 M Tris-HCl, 0.1 M NaCl

As far as the effect of temperature is concerned, the intrinsic temperature sensitivity is extremely high for both whale hemoglobins, as shown in Table 1; when the physiological cofactors are added to the two systems, the overall heat required for oxygenation of *B. acutorostrata* Hb, after correction for the heat of oxygen solubilization, ranges from -3.5 to -4.0 kcal/mol of oxygen, whereas *B. physalus* hemoglobin displays ΔH values that range from -12.3 to -12.5 kcal/mol.

Table 1. Heat of oxygenation (ΔH) in kcal/mol of whale hemoglobin (in 0.1 M Tris-HCl, 0.1 M NaCl, pH 7.5)

Species	Stripped	+1% CO_2	+2,3-DPG
Balaenoptera physalus	-15.57	-12.3	-12.6
Balaenoptera acutorostrata	-12.0	-3.5	-4.0

Discussion

The results obtained show the peculiar temperature-dependent effect of carbon dioxide on these two whale hemoglobins. This interplay of temperature and carbon dioxide is of great physiological importance and has been explained on the basis of the lower temperature that whales experience at the level of the fins and tail, which are sites of great muscular activity. In fact, within the core of the body, carbon dioxide does not display any allosteric effect because at 37°C there is no differential binding of this ligand with respect to oxy- and deoxy-structure. However, at the fins and tail, where the temperature is always below 37°C (with precise values depending on sea water temperature), carbon dioxide facilitates oxygen unloading as it does in humans. The fact that this behavior is shared by both whale hemoglobins underlines the general validity of this mechanism for large diving mammals. As regards the different temperature sensitivity shown by the two species of whales, this may be explained by the different habitats in which they live. *B. acutorostrata*, which lives in the Arctic sea, is subjected to very low environmental temperatures and so its hemoglobin needs to have low temperature sensitivity to optimize oxygen unloading (as all Arctic mammals do); whereas *B. physalus*, which lives in the Mediterranean sea where extremely low temperatures are never reached, does not require hemoglobin with a low temperature sensitivity. These results are of particular evolutionary interest and underline the role of temperature and its interplay with certain heterotropic ligands in the modulation of respiratory proteins in relation to habitat.

References

1. di Prisco G, Condò, SG, Tamburrini M, Giardina B (1991) Oxygen transport in extreme environments. Trends Biochem Sci 16:471-474
2. Lykkeboe G, Johansen K, Maloy GMO (1975) Functional properties of hemoglobin in the teleost *Tilapia grahami*. J Comp Physiol 104:1-11
3. Sick H, Gersonde K (1969) Method for continuous registration of O_2-binding curves of hemoproteins by means of a diffusion chamber. Anal Biochem 32:362-376
4. Dhindsa DS, Metcalfe J, Hoversland AS, Hautman RA (1974) Comparative studies of the respiratory functions of mammalian blood. X. Killer whale (*Orcinus orca* Linneus) and brown galago (*Galago crassicaudatus crassicaudatus*). Resp Physiol 20:93-103
5. Kilmartin JV, Rossi Bernardi L (1973) Interaction of hemoglobin with hydrogen ions, carbon dioxide and organic phosphates. Physiol Rev 53:836-890
6. Jelkmann W, Oberthur W, Kleinschmidt T, Braunitzer G (1981) Adaptation of hemoglobin function to subterranean life in the mole, *Talpa europea*. Resp Physiol 46: 7-16

Oxygen Transport and Diving Behaviour: The Haemoglobin from Dolphin Tursiops truncatus

E. Tellone[1], M.E. Clementi[2], A.M. Russo[1], S. Ficarra[1], A. Lania[1], A. Lupi[2], B. Giardina[2], A. Galtieri[1]

Introduction

Haemoglobins (Hb) in their purified state exhibit a great deal of variation in terms of absolute affinities for oxygen and their susceptibility to control by effectors such as chloride, carbon dioxide, protons and organic phosphates, namely, 2,3-diphosphoglycerate (DPG). This is generally thought to be a reflection of both the variable oxygen tension in which organisms live and the variable oxygen demands of their respiring tissues.

The modulation of function induced by all the effectors mentioned above has important physiological effects. As an example, at the level of tissues, the decrease of oxygen affinity brought about by the increase in proton activity (alkaline Bohr effect) allows a more efficient unloading of oxygen and contributes to the neutralization of proton produced by CO_2 and lactic acid.

Another important feature, which should not be disregarded, of the reaction of Hb with O_2 is its temperature dependence, which is determined by the associated overall enthalpy change (ΔH). In this respect, it may be worthwhile to point out that the overall enthalpy change of O_2 binding is the result of different contributions which may be summarized as follows: (1) intrinsic heat of oxygenation, namely the heat involved in the binding of O_2 to the haeme iron; (2) heat of ionization of O_2-linked ionizable groups (Bohr groups), which is always endothermic, i.e. ΔH positive; (3) heat associated with the $T \rightarrow R$ allosteric transition; and (4) heat of binding of other ions such as organic phosphates and chloride ions.

[1]Department of Organic and Biological Chemistry, University of Messina, Italy
[2]Institute of Chemistry, Faculty of Medicine and CNR Center for Receptors Chemistry, Catholic University, Rome, Italy

G. di Prisco, B. Giardina, R.E. Weber (Eds)
Hemoglobin Function in Vertebrates.
Molecular Adaptation in Extreme and Temperate Environments
© Springer-Verlag Italia 2000

From the aforementioned considerations, it appears clear that the overall effect of temperature could be modulated through the linkage between the basic reaction (i.e. the oxygenation-deoxygenation cycle) and the binding of different ions and effectors.

Hence, a number of studies [1, 2] of Hb function in different species at varying temperatures have revealed a series of adaptive mechanisms which are based on the thermodynamic connection between the binding of heterotropic effectors and the reaction with oxygen.

From the physiological point of view it may be important to outline that this thermodynamic approach is of great interest not only in the case of cold-blooded species such as fish, but also in mammalian species where temperature gradients of 10°-15°C commonly occur between the core of the organism and its periphery [3, 4].

Along this line we liked to investigate the functional properties of the Hb from the dolphin *Tursiops truncatus*. In fact, due to the characteristic diving behaviour of dolphins, the blood of these animals has to accomplish its oxygen transport function under a wide range of conditions facing marked variations in pH levels and substantial temperature changes.

The whole body of the data outlines once more the physiological significance of a thermodynamic analysis of the reaction with oxygen and of all the O_2-linked heterotropic effects.

Material and Methods

Dolphin (*Tursiops truncatus*) blood samples were obtained from animals bred in the aquarium in Riccione (Italy) and fresh human blood samples from a blood bank. The blood samples were collected in the presence of heparin. The red cells were washed three times with iso-osmotic NaCl solution by centrifugation at 1000 g, and the packed cells lysed by adding 2 volumes of cold hypotonic buffer. The stroma was removed by centrifugation at 12,000 g for 30 min. Electrophoretic analysis was performed by alkaline polyacrylamide gel electrophoresis.

Stripped Hb was obtained by first passing the haemolysate through a Sephadex G-25 column equilibrated with 0.01 M tris buffer, pH=8, containing 0.1 M NaCl and then through a column of mixed-bed ion-exchange resin (Bio-Rad AG 501_8).

Concentrated stock solutions of DPG were prepared by dissolving the sodium salt of 2,3-diphosphoglyceric acid (Sigma) in Hepes buffer.

Oxygen-binding isotherms were determined by the tonometric method, in the absence and presence of allosteric effectors, between 15°C and 37°C. The overall oxygenation enthalpy corrected for the solubilization heat of O_2 (ΔH=-3 kcal·mol^{-1}) was calculated from the integrated van't Hoff equation:

$$\Delta H = -4.574 \cdot [(T_1 \cdot T_2) / (T_1 - T_2)] \cdot \Delta log P_{50} / 1000, \text{ kcal·mol}^{-1},$$

where P_{50} is the partial pressure of the ligand at which 50% of haemes is oxygenated. Over the temperature range explored (20°C-40°C) Van't Hoff plots were linear within the experimental error. An average standard deviation of ±8% for values of P_{50} and of ±15% for ΔH values was calculated.

Under all the experimental conditions, the ferric derivative of dolphin Hb (met-Hb) was always lower than 3%.

Results

Electrophoretic analysis of the haemolysate from *Tursiops truncatus* has indicated the presence of only one Hb component, characterized by a mobility very similar to that of human HbA (data not shown).

Figure 1 shows the Bohr effect of dolphin Hb investigated within the pH range 6.3-8.2 both in the absence and presence of 2,3-DPG, at 20° and 37°C. As evident at 20°C and in the absence of DPG, dolphin Hb is characterized by a Bohr effect similar in amplitude to that displayed by human Hb (see Fig. 2). The situation is different at 37°C and in the presence of DPG. In fact, the Bohr coefficient ($\Delta logP50/\Delta pH$), calculated only in the presence of chloride ions, is 0.42 for dolphin Hb and 0.46 for HbA at 20°C and becomes, at 37°C, 0.55 and 0.43 for dolphin and human Hb, respectively. In the presence of DPG, the Bohr coefficient is always higher for dolphin Hb respect to that of human Hb, with values of 0.55 versus 0.43 at 20°C and 0.83 versus 0.67 at 37°C.

The results reported in Fig. 3 shows the effects of increasing concentrations of 2,3-DPG on the oxygen affinity of dolphin Hb in Hepes buffer plus 0.1 M NaCl at pH 6.5 and 20°C. In the same figure, the data relative to human Hb, under a similar set of conditions, are also reported. In the absence of organic phosphates the oxygen affinity for oxygen of human Hb is higher than that of dolphin Hb, and this difference also remains constant in the presence of different concentrations of DPG.

However, a peculiar characteristic of dolphin Hb concerns its response to changes in temperature; Fig. 4 shows the effect of temperature and the influence of pH on the overall oxygenation enthalpy (ΔH) for *Tursiops truncatus* Hb both in presence and absence of 2,3-DPG. At pH 7.5 the overall ΔH obtained under stripped conditions (without chloride ions and DPG) is -10.7 kcal/mol (data not shown), demonstrating a strong exothermic reaction similar to that in human Hb. When the experiments were carried out in the presence of chloride ions and DPG, the temperature effect is markedly decreased to about one half of the original effect. It should also be noted that the ΔH values are negative at neutral and alkaline pH values but become positive at acid pH, being +1.2 kcal/mol of oxygen at pH 6.5 in the presence of organic phosphates.

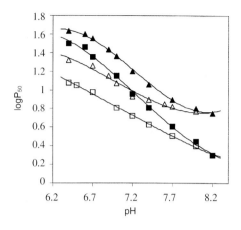

Fig. 1. Bohr effect of dolphin haemoglobin in the absence (*open symbols*) and presence (*closed symbols*) of 3 mM 2,3-DPG at 20C° (*squares*) and 37°C (*triangles*). Conditions: 0.1 M Hepes buffer plus NaCl 0.1 M

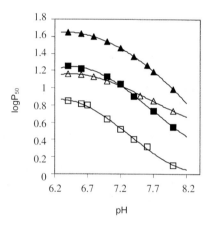

Fig. 2. Bohr effect of human haemoglobin in the absence (*open symbols*) and presence (*closed symbols*) of 3 mM 2,3-DPG at 20°C (*squares*) and 37°C (*triangles*). Conditions: 0.1 M Hepes buffer plus NaCl 0.1 M

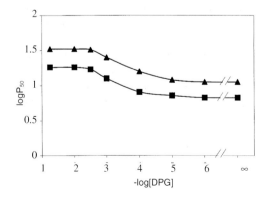

Fig. 3. Effect of 2,3-DPG concentration on the oxygen affinity of dolphin (*triangles*) and human (*squares*) haemoglobin. Conditions: 0.1 M Hepes buffer NaCl 0.1 M pH 6.5, 20°C

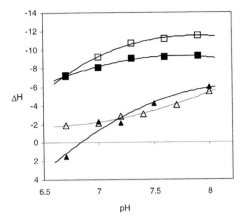

Fig. 4. Overall ΔH values (expressed in kcal·mol⁻¹ of oxygen) as a function of pH for dolphin (*triangles*) and human (*squares*) haemoglobin in the absence (*open symbols*) and in the presence (*filled symbols*) of 3 mM 2,3-DPG. The values are calculated from van't Hoff equation, by using the data obtained from oxygen equilibria experiments and are corrected for the heat contribution of oxygen in solution (-3 kcal·mol⁻¹). Conditions: 0.1 M Hepes buffer plus 0.1 *M* NaCl

Discussion

The oxygen-binding behaviour of dolphin Hb has revealed some properties which are probably linked to the peculiar diving behaviours of this animal. In fact, dolphin Hb is characterized by a Bohr effect similar to that in human Hb but with a substantial shift of the midpoint to the transition towards acidic values (midpoint values are 7.0 and 7.7 for dolphin and human Hb, respectively); this functional characteristic is necessary to maintain an efficient unloading of oxygen to tissues during the prolonged apnoea that this mammal performs while diving to considerable depths.

A thermodynamic analysis of *Tursiops truncatus* Hb shows a strong variation in the overall heat of oxygenation, measured in the pH range between 6.5 and 7.5. A more exothermic value of ΔH is observed at alkaline pH values, but approaching the acidic range, the overall heat of oxygenation becomes less and less exothermic.

This strong variation in enthalpy is probably due to the increasing endothermic contribution of the Bohr protons, which cancels some of the heat released upon O_2 binding. In looking for the possible physiological significance of this phenomenon, it seems particularly important that the overall ΔH is positive precisely at those pH values which dolphin tissues reach during prolonged dives due to the decrease in pH brought about by lactic acid production. Under these conditions the tissues will have a more acidic pH that will lower the oxygen affinity of the Hb and render the ΔH of oxygenation process positive, promoting the unloading of oxygen when the animal encounters colder water during diving.

Summary

The oxygen-binding properties of the single Hb from the dolphin *Tursiops truncatus* have been investigated as a function of protons, chloride ions, organic phosphates and temperature.

We have found that this Hb shows a Bohr effect which is similar in amplitude but characterized by a shift of the midpoint towards acidic values as compared to human Hb.

Moreover, the heat of oxygenation, expressed by ΔH values, becomes positive at lower pH. These features may be of great physiological importance because they seem to be well adapted for gas exchange during prolonged dives, which are often taken in cold waters. In particular, if the shift of the Bohr effect towards acid pH values is related to the diving behaviour of the animal, the same might apply to the endothermic ΔH of oxygen binding at lower pH that could be linked to the necessity of promoting the unloading of oxygen at the level of fins and tail in cold environments.

References

1. di Prisco G, Condò SG, Tamburrini M, Giardina B (1991) Oxygen transport in extreme enviroments. Trends Biochem. Sci. 16: 471-474
2. Giardina B, Galtieri A, Lania A, Ascenzi P, Desideri A, Cerroni L, Condo SG (1992) Reduced sensitivity of O2 transport to allosteric effectors and temperature in loggerhead sea turtle hemoglobin: functional and spectroscopic study. Biochim Biophys Acta 1159(2):129-133
3. Coletta M, Clementi ME, Ascenzi P, Petruzzelli R, Condo SG, Giardina B (1992) A comparative study of the temperature dependence of the oxygen-binding properties of mammalian hemoglobins. Eur J Biochem 204(3):1155-1157
4. Giardina B, Scatena R, Clementi ME, Cerroni L, Nuutinen M, Brix O, Sletten SN, Castagnola M, Condo SG (1993) Physiological relevance of the overall ΔH of oxygen binding to fetal human hemoglobin J Mol Biol; 229(2):512-516

Molecular Modelling Analysis of the Haemoglobins of the Antarctic Bird *Catharacta maccormicki*: the Hypothesis of a Second Phosphate Binding Site

A. Riccio[1], M. Tamburrini[1], B. Giardina[2], G. di Prisco[1]

Introduction

Antarctic organisms are exposed to very low temperatures. Thus, in order to face extreme life conditions, suitable mechanisms of cold adaptation have been developed, involving physiological and biochemical processes [1].

In the framework of a study on the oxygen-transport and storage systems of Antarctic marine organisms, we began to investigate adaptations of such systems in Antarctic birds. This research is part of a series of investigations which have already provided insight into the role of temperature and of its interplay with heterotropic ligands in the modulation of haemoglobin (Hb) function [2-12].

We have already reported studies on the structure/function relationship in Emperor penguin (*Aptenodytes forsteri*) Hb and myoglobin (Mb), in search of correlation with the bird's life style [13, 14]. The functional properties of both haemoproteins seem well adapted for gas exchange during prolonged dives. The weak alkaline Bohr effect preserves penguin Hb from a sudden and uncontrolled stripping of oxygen. Moreover, the very minor enthalpy change observed at low pH values may be considered an example of molecular adaptation, through which oxygen delivery becomes essentially insensitive to exposure to the extremely low temperatures of the environment. Addition of lactate has no major effect on Mb oxygenation over a wide temperature range; therefore, the Mb function is not affected by metabolic acidosis induced by prolonged physical effort such as diving.

[1]Institute of Protein Biochemistry and Enzymology, CNR, Naples, Italy
[2]Institute of Chemistry and Clinical Chemistry, Faculty of Medicine, Catholic University "Sacro Cuore", Rome, Italy

G. di Prisco, B. Giardina, R.E. Weber (Eds)
Hemoglobin Function in Vertebrates.
Molecular Adaptation in Extreme and Temperate Environments
© Springer-Verlag Italia 2000

Here we report on an investigation of the two Hbs of south polar skua (*Catharacta maccormicki*, Stercorariidae), a seabird breeding in coastal Antarctic regions. Skuas are well known for their aggressive rapacious habits. They have a rapid, sustained and powerful flight, enabling them to overtake many birds. Skuas also prey on chicks and eggs, particularly those of penguins. Adults stay near breeding colonies in the Antarctic during the summer, but move north to subtropical waters during the winter, occasionally reaching the northern hemisphere [15]. The Hbs showed peculiar functional features, probably acquired to meet special needs in relation to the environmental conditions. Similar to Emperor penguin Hb, both skua Hbs showed a weak, alkaline Bohr effect which may impair excessive oxygen delivery in response to acidosis which, during prolonged flight, may occur due to the great muscular activity [16]. Moreover, the experimental evidence showed that both skua Hbs have two functionally operative phosphate binding sites per tetramer. For the first time molecular modelling analysis strongly supported the hypothesis at the structural level of a second binding site for polyphosphates, suggesting a possible physiological role for this site in skua Hbs. The hypothesis that most Hbs possess a potential second binding site is also discussed.

The Second Binding Site for Polyphosphates

The existence of more than one binding site for polyanions has been hypothesised in bovine [17, 18] and horse [19] Hbs. In particular, it was found that β-naphthyl triphosphate (β-NapP$_3$) binds to human deoxy-Hb A tetramer in the molar ratio 2:1, and that one of the two β-NapP$_3$ competes with 2,3-diphosphoglyceric acid (2,3-DPG) [18].

Studies on human Hb [20] proposed a second site and suggested that it might be formed by the cluster of the four positive charges of the N and C termini of the two α chains. This hypothesis was strongly supported by the observed decrease of phosphate affinity after carbamylation of the N-terminal residues of the α chains. Furthermore, having observed that the fast exchange between bound and free inositol hexakisphosphate (IHP) on the NMR time scale cannot be reconciled with a single-step binding mechanism, the introduction of an entry-leaving site for phosphate was required. Moreover, studies on the binding of IHP to the oxygenated, carbonylated and nitrosylated derivatives of ferrous human Hb demonstrated that binding of phosphate is characterised by a byphasic profile, suggesting the presence of at least two IHP binding sites per tetramer with different affinity constants [21].

Binding of IHP to the oxygenated derivative of dromedary (*Camelus dromedarius*) and human Hb showed that dromedary Hb binds two IHP molecules per tetramer at distinct sites with different affinities, whereas human Hb binds only one IHP molecule per tetramer [22].

The hypothesis of the second site is also supported by the observation that in dromedary Hb P_{50} (the partial pressure of oxygen required to saturate 50%

of the molecules) clearly exhibits a biphasic character when investigated in titration curves with 2,3-DPG [23]. This effect indicates that there must be at least two ligand-linked binding sites for phosphates, one of which is made stronger and the other weaker, in terms of affinity, as a result of Hb oxygenation. Further studies provided additional experimental evidence for the existence in dromedary Hb of two polyanion binding sites which affect the conformational equilibrium of the macromolecule in an opposite way [24].

It has also been shown that in bear (*Ursus arctos*) Hb there are two classes of chloride-binding sites, one acting synergistically with 2,3-DPG and another one probably overlapping with the organic phosphate binding site, which is therefore fully operative only in the absence of 2,3-DPG [25]. Moreover, the two heterotropic binding sites have direct and separate communication pathways with the haeme.

We extended the aforementioned studies to the interactions between phosphate and skua Hbs, taking detailed molecular models into account for the first time and highlighting the fundamental role of Lys99α in the structure of the second binding site.

Skua Hbs

The haemolysate of adult *C. maccormicki* contains two major components (Hb 1 and Hb 2), accounting approximately for 65% and 35% of the total.

Phosphate binding was investigated using IHP. This effector is structurally and functionally very similar to inositol pentakisphosphate (IPP), the physiological allosteric co-factor of bird Hbs. The effect of IPP is actually only slightly lower than that of IHP [26-29].

The titration curves of the oxygen affinity of the two Hbs as a function of chloride and phosphate concentration indicated a very strong effect of phosphate, which is almost constant in the physiological pH range. The curves obtained with increasing concentrations of phosphate showed a double plateau both in the absence and presence of saturating concentrations of chloride. This evidence suggested the presence of two distinct polyanion binding sites, characterised by different affinities for IHP.

Since detailed analysis at the molecular level of the phosphate binding sites requires the deoxy T-state structure, and since crystallographic structures of avian deoxy Hbs are not available, the high-resolution structure of human deoxy Hb [30] was used as template to build the molecular models of skua Hbs.

Computer graphics, structural manipulations, energy minimisation calculations and molecular dynamic simulations were carried out with a Silicon Graphics Indigo^2 workstation using the Insight II software package (Biosym/Molecular Simulation Incorporated, MSI). Two IHP molecules were manually added to the two skua Hbs, the first one in the main binding site, i.e. the central cavity located between the two β chains of the deoxy form [31], and the second one in the cavity formed by the N and C termini of the two α chains, previously suggested to be a second site in other Hbs [20, 23]. The human Hb A-(IHP)$_2$ molecular complex was built and refined with the same procedure and used as control to ascertain the validity of the results obtained with skua Hbs.

An analysis of the molecular models of skua Hb 1 and Hb 2 revealed that the main phosphate binding site in skua Hbs – which have the β chain in common - is made of seven residues of each β chain (sequence positions 1, 2, 82, 135, 136, 139, 143), which are in direct contact with IHP.

The second site, located between the two α chains in both Hbs, is made of seven residues of each α chain, instead of the two (the N and C termini) suggested in other Hbs [20, 23]. These residues (sequence positions 1, 95, 99, 134, 137, 138, 141) are in direct contact with IHP (Fig. 1). The interaction of IHP with skua and human Hbs in this region is enhanced by several hydrogen bonds, as well as by strong salt bridges established with Lys99α. It is noteworthy that, although skua Hbs have different α chains, and there are four differences among the residues forming their additional phosphate binding sites (Table 1), Lys99α is conserved in Hb 1 and Hb 2.

Table 1. α-Chain residues forming the additional IHP binding sites of skua Hbs and human Hb A

skua Hb 1	Val1	Pro95	Lys99	Asn134	Thr137	Ala138	Arg141
skua Hb 2	Met1	Pro95	Lys99	Ala134	Ser137	Glu138	Arg141
human Hb A	Val1	Pro95	Lys99	Thr134	Thr137	Ser138	Arg141

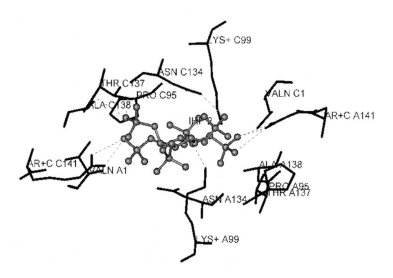

Fig. 1. Additional phosphate-binding site of skua Hb 1, after minimisation. IHP, represented as *ball and stick*, is shown in *grey*. The H-bonds established between IHP and Hb are indicated by *grey dashed lines*. The residues indicated with *A* and *C* belong to the α$_1$ and α$_2$ chains, respectively

Discussion

A second binding site for polyphosphates has previously been suggested in several mammal Hbs on the basis of different experimental approaches [17-25]. The site has tentatively been proposed to be formed by the cluster of the four positive charges of the N and C termini of two α chains [20, 23].

This investigation was undertaken as part of a study of oxygen transport and release in Antarctic birds and analyses the molecular models of skua Hb 1 and Hb 2 using human Hb A as a template. The IHP complexes were studied, analysing the interactions of phosphate with the main binding site, located between the two β chains, and with the suggested second site, located between the two α chains.

The results indicate that there are twelve charged groups forming the main site of skua Hbs instead of the eight of Hb A [31], whereas six groups form the second site instead of the four so far reported in other Hbs [20,23], which overlooks the contribution of Lys99α. Therefore, our model can explain why in Hb A carbamylation of the N termini of the α chains does not completely abolish the affinity of this site for IHP [20]. In fact, carbamylation leaves four positively charged groups able to bind IHP.

The amino acid residues forming both sites in skua Hb 1 and Hb 2 are mostly conserved in comparison with other Antarctic avian major and minor Hbs, respectively [13; Tamburrini et al. unpublished]. Therefore, the structure of the two phosphate binding sites of skua Hbs is likely to be of general significance, warranting further investigation of the physiological relevance in other birds.

Under physiological conditions, the second site would enhance the ability of Hb to capture phosphate from the solution and readily transfer it to the main site by means of a site-site migratory mechanism, thus acting as an entry-leaving site and favouring the release of oxygen. This model is in agreement with that proposed by Zuiderweg et al. [20], according to which phosphate is able to migrate on the same Hb molecule between the two binding sites.

The second phosphate-binding site may also contribute to over-stabilise the T structure, further favouring oxygen release. This stabilising effect, although lower than that due to phosphate binding between the β chains, could be helpful for long-distance migrations, especially when physiological stress becomes an important factor.

In summary, this investigation strongly supports the existence of a second, lower affinity binding site for polyphosphates, analysed for the first time at the structural level on the molecular models of skua Hbs. The observation that Lys99α (whose contribution in the structure of the second site has so far been overlooked) is strongly conserved suggests that most Hbs possess a potential second site, whose physiological relevance may well be linked to evolutionary adaptations and life style.

Acknowledgements

This study was conducted as part of the Italian National Programme for Antarctic Research.

References

1. di Prisco G (1997) Physiological and biochemical adaptations in fish to a cold marine environment. In: Battaglia B, Valencia J, Walton DWH (eds) Antarctic communities: species, structure and survival. Proc SCAR 6th Biol Symp, Venice. Cambridge University Press, Cambridge, pp 251-260

2. Brix O, Bardgard A, Mathisen S, El-Sherbini S, Condò SG, Giardina B (1989) Arctic life adaptation. I. The function of musk ox (*Ovibos muschatos*) haemoglobin. Comp Biochem Physiol 94B:135-138

3. Brix O, Condò SG, Bardgard A, Tavazzi B, Giardina B (1990) Temperature modulation of oxygen transport in a diving mammal (*Balaenoptera acutorostrata*). Biochem J 271:509-513

4. Giardina B, Brix O, Nuutinen M, El-Sherbini S, Bardgard A, Lazzarino G, Condò SG (1989) Arctic adaptation in reindeer. The energy saving of haemoglobin. FEBS Lett 247:135-138

5. Giardina B, Condò SG, Petruzzelli R, Bardgard A, Brix O (1990) Thermodynamics of oxygen binding to arctic haemoglobins. The case of reindeer. Biophys Chem 37:281-286

6. Giardina B, Condò SG, Brix O (1992) The interplay of temperature and protons in the modulation of oxygen binding to squid blood. Biochem J 281:725-728

7. Giardina B, Galtieri A, Lania A, Ascenzi P, Desideri A, Cerroni L, Condò SG (1992) Reduced sensitivity of oxygen transport to allosteric effectors and temperature in loggerhead sea turtle haemoglobin: functional and spectroscopic study. Biochim Biophys Acta 1159:129-133

8. Giardina B, Ascenzi P, Clementi ME, De Sanctis G, Rizzi M, Coletta M (1996) Functional modulation by lactate of myoglobin. A monomeric allosteric hemoprotein. J Biol Chem 271:16999-17001

9. di Prisco G (1998) Molecular adaptations of Antarctic fish hemoglobins. In: di Prisco G, Pisano E, Clarke A (eds) Fishes of Antarctica. A biological overview. Springer-Verlag Italia, Milano, pp 339-353

10. di Prisco G, Giardina B (1996) Temperature adaptation: molecular aspects. In: Johnston IA, Bennett AF (eds) Animals and temperature. Phenotypic and evolutionary adaptation. Soc Exptl Biol, Seminar Series 59, Cambridge, Cambridge University Press, pp 23-51

11. di Prisco G, Condò SG, Tamburrini M, Giardina B (1991) Oxygen transport in extreme environments. Trends Biochem Sci 16:471-474

12. di Prisco G, D'Avino R, Tamburrini M (1999) Structure and function of hemoglobins from Antarctic organisms: the search for correlations with adaptive evolution. In: Margesin R, Schinner F (eds) Cold-adapted organisms. Ecology, physiology, enzymology and molecular biology. Springer, Berlin Heidelberg New York, pp 239-253

13. Tamburrini M, Condò SG, di Prisco G, Giardina B (1994) Adaptation to extreme environments: structure-function relationships in Emperor penguin haemoglobin. J Mol Biol 237:615-621

14. Tamburrini M, Romano M, Giardina B, di Prisco G (1999) The myoglobin of Emperor penguin (*Aptenodytes forsteri*): amino acid sequence and functional adaptation to extreme conditions. Comp Biochem Physiol 122B:235-240

15. Watson GE (1975) Birds of the Antarctic and Sub-Antarctic. Antarctic Research Series. William Byrd, Richmond

16. Giardina B, Corda M, Pellegrini MG, Sanna MT, Brix O, Clementi ME, Condò SG (1990) Flight and heath dissipation in birds. A possible molecular mechanism. FEBS Lett 270:173-176

17. Kuwajima T, Asai H (1975) Synthesis of fluorescent organic phosphates and their equilibrium binding to bovine oxyhemoglobin. Biochemistry 14:492-497

18. Horiuchi K, Asai H (1983) Binding of β-naphthyl triphosphate to human adult hemoglobin accompanying deoxygenation, investigated by simultaneous measurements of fluorescence, absorbance and partial pressure of oxygen. Eur J Biochem 131:613-618

19. Hedlund B, Danielson C, Lovrien R (1972) Equilibria of organic phosphates with horse oxyhemoglobin. Biochemistry 11:4660-4668

20. Zuiderweg ERP, Hamers LF, Rollema HS, de Bruin SH, Hilbers CW (1981) ^{31}P NMR study of the kinetics of binding of *myo*-inositol hexakisphosphate to human hemoglobin. Eur J Biochem 118:95-104

21. Coletta M, Ascenzi P, Santucci R, Bertollini A, Amiconi G (1993) Interaction of inositol hexakisphosphate with liganded ferrous human hemoglobin. Direct evidence for two functionally operative binding sites. Biochim Biophys Acta 1162:309-314

22. Ascenzi P, Amiconi G, Rossi E, Segre AL (1989) Binding of inositol hexakisphosphate to the oxygenated derivative of dromedary (*Camelus dromedarius*) and human hemoglobin: [31]p-NMR study. J Inorg Biochem 35(4):247-253

23. Amiconi G, Bertollini A, Bellelli A, Coletta M, Condò SG, Brunori M (1985) Evidence for two oxygen-linked binding sites for polyanions in dromedary hemoglobin. Eur J Biochem 150:387-393

24. Desideri A, Ascenzi P, Chiancone E, Amiconi G (1987) Effect of inositol hexakisphosphate on the EPR properties of the nitric oxide derivative of ferrous dromedary (*Camelus dromedarius*) hemoglobin. Evidence for two polyanion binding sites. J Inorg Biochem 29(2):131-135

25. Coletta M, Condò SG, Scatena R, Clementi ME, Baroni S, Sletten SN, Brix O, Giardina B (1994) Synergistic modulation by chloride and organic phosphates of hemoglobin from bear (*Ursus arctos*). J Mol Biol 236:1401-1406

26. Brygier J, Paul C (1976) Oxygen equilibrium of chicken haemoglobin in the presence of organic phosphates. Biochemie 58:755-756

27. Vandecasserie C, Fraboni A, Schnek AG, Leonis J (1976) Oxygen affinity of some avian haemoglobins in presence of various phosphorilated cofactors. Colloque sur l'hemoglobine. Le Touquet-Paris-Plage, p 34

28. Lutz PL (1980) On the oxygen affinity of bird blood. Am Zool 20:187-198

29. Giardina B, Corda M, Pellegrini MG, Condò SG, Brunori M (1985) Functional properties of the haemoglobin system of two diving birds (*Podiceps nigricollis* and *Phalacrocorax carbo sinensis*). Mol Physiol 7:281-292

30. Fermi G, Perutz MF, Shaanan B, Fourme R (1984) The crystal structure of human deoxyhaemoglobin at 1.74 Å resolution. J Mol Biol 175:159-174

31. Arnone A, Perutz MF (1974) Structure of inositol hexaphosphate-human deoxyhemoglobin complex. Nature 249:34-36

Transport of Oxygen during Hibernation: the Hemoglobin of *Dryomys nitedula*

M.E. Clementi[1], S. Ficarra[2], A. Galtieri[2], A. Lupi[1], B. Giardina[1]

Introduction

Oxygen transport proteins might well represent an interesting model which can be used to evaluate the different strategies adopted by evolution to solve the problem of oxygen supply in accordance with the physiological requirements of the specific tissues of a given organism. This is clearly demonstrated by various mammalian hemoglobins which have developed a number of complex regulatory mechanisms involving binding with allosteric effectors (such as organic phosphates, CO_2, chloride ions and Bohr protons) and their interplay with temperature.

As regards temperature, it has been generally assumed that mammalian blood has a greater temperature sensitivity than that of ectothermic organisms; however, recent results have shown that, in some species of mammals, hemoglobin displays overall oxygenation enthalpy (ΔH) values which are much less exothermic than those observed for human adult hemoglobin [1, 2].

Therefore, it seemed of particular interest to investigate the functional properties of the hemoglobin of *Dryomys nitedula* (termed forest dormouse), a hibernating animal, to study the effect of temperature and physiological effectors on oxygen transport: Hibernation represents physiological conditions characterized by a hypometabolic state in which the body temperature decreases from ca. 35-37°C in the active state to ca. 5°C. In addition, during this period the 2,3-BPG concentration in red cells and the intracellular pH also decrease significantly [3-5].

[1]Institute of Chemistry, Faculty of Medicine and CNR Center for Receptors Chemistry, Catholic University, Rome, Italy
[2]Department of Organic and Biological Chemistry, University of Messina, Italy

G. di Prisco, B. Giardina, R.E. Weber (Eds)
Hemoglobin Function in Vertebrates.
Molecular Adaptation in Extreme and Temperate Environments
© Springer-Verlag Italia 2000

The results of these studies seem of particular interest, since they show that, in spite of the physiological modifications that occur during hibernation (hypothermia, acidosis, and lowering of 2,3-BPG concentration), the hemoglobin of *Dryomys nitedula* allows an efficient oxygenation-deoxygenation cycle for the basal functions.

Materials and Methods

Blood samples from animals were collected into an isotonic NaCl solution containing 2 mM EDTA. The cells were washed three times with iso-osmotic NaCl solution by centrifugation at 1000 g and the packed cells lysed by adding 2 volumes of cold hypotonic buffer. The stroma was removed by centrifugation at 12,000 g for 30 min.

Chloride and organic phosphates were removed (stripped hemoglobin) by first passing the hemolysate through a Sephadex G-25 column, equilibrated with 0.01 M Tris-HCl buffer pH 8.0 containing 0.1 M NaCl, and then through a column of mixed-bed ion-exchange : Concentrated stock solutions of BPG were prepared by dissolving the sodium salt of 2,3-biphosphoglyceric acid in Hepes buffer.

Oxygen-binding isotherms were determined by the tonometric method [6] at a protein concentration of 3-5 mg/ml, in the absence and presence of allosteric effectors, between 5°C and 37°C. The overall oxygenation enthalpy, corrected for the solubilization heat of O_2 (ΔH=-3 kcal·mol^{-1}), was calculated according to the integrated van't Hoff equation:

$$\Delta H = -4.574 \cdot [(T_1 \cdot T_2) / (T_1 - T_2)] \cdot \Delta \log P_{50} / 1000, \text{kcal·mol}^{-1},$$

where P_{50} is the partial pressure of the ligand at which 50% of the hemes are oxygenated. Over the temperature range explored the van't Hoff plots were linear within the experimental error. An average standard deviation of ±8% for values of P_{50} and of ±15% for ΔH values was calculated.

Curve fitting as a function of effector concentration was carried out on a PDP II/23 (Digital Equipment, USA) using a non-linear least-squares fitting procedure with a Marquardt algorithm according to the following equation:

$$\log P_{50}^{obs} = \log P_{50}^0 + R \log \{(1 + K_d [E]) / (1 + K_o [E])\}$$

where P_{50}^{obs} refers to the oxygen affinity observed at a given concentration [E] of the effector under investigation, P_{50}^0 corresponds to the oxygen affinity displayed in the absence of the effector, K_d and K_o are the association equilibrium constants for the effector to unliganded and liganded hemoglobin, respectively, and R corresponds to the number of effector binding sites per heme (i.e., R=0.25 for DPG).

Results

Figures 1 and 2 show the effect of pH on the oxygen affinity of dormouse hemoglobin in the presence and absence of both chloride ions and BPG at 5° and 37°C, respectively. As expected, the presence of chloride and organic phosphates decreases at both the temperatures, the oxygen affinity of this hemoglobin. However, in contrast to human hemoglobin in which the presence of the allosteric effectors increases the Bohr coefficient (data not shown), the Bohr effect has the same amplitude under all the conditions studied. Moreover, this is evident in comparison to the data obtained from human hemoglobin (reported in Table 1), as the functional properties of *Dryomys* hemoglobin are only weakly affected by the presence of the various effectors and, in particular, by the decrease in temperature. In fact, whereas a temperature of 5°C dramatically increases the oxygen affinity of human hemoglobin up to values of P_{50} equalling 2.2 mmHg (at pH=7.4 and in the presence of chloride ions and BPG at physiological concentrations), dormouse hemoglobin, under the same conditions, may be able to provide the oxygen unloading and loading functions (P_{50}=11.22 mmHg).

Table 1. Oxygen affinity (expressed as $logP_{50}$) for human hemoglobin under different experimental conditions at 5° and 37°C. The experiments were performed in Hepes buffer 100 mM in the presence of 100 mM chloride ions and 3 mM BPG

	$logP_{50}$ (5°C)	$logP_{50}$ (37°C)
NaCl		
pH 6.8	0.20	1.15
pH 7.4	-0.15	0.96
plus NaCl and BPG		
pH 6.8	0.75	1.60
pH 7.4	0.35	1.36

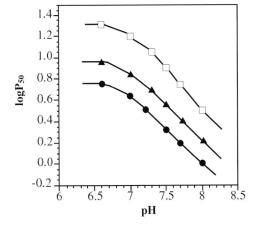

Fig. 1. Bohr effect for *Dryomys nitedula* hemoglobin in the absence (*circles*) and presence (*triangles*) of 0.1 M chloride and presence (*squares*) of 3 mM 2,3-BPG. Conditions: Mes or Hepes or Taps 0.1 M at 5°C

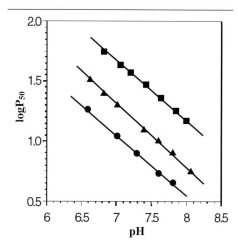

Fig. 2. Bohr effect for *Dryomys nitedula* hemoglobin in the absence (*circles*) and presence (*triangles*) of 0.1 *M* chloride and in the presence (*squares*) of 3 m*M* 2,3-BPG. Conditions: Mes or Hepes or Taps 0.1 *M* at 37°C

Furthermore, this is also shown by the presence of only chloride ions, which minimize the effect of temperature on the reduction in dormouse hemoglobin affinity. In Fig. 3 the overall ΔH values in the function of pH are reported for dormouse and human hemoglobins in the presence and absence of BPG. As is evident, the hemoglobin from hibernating animal shows a very low sensitivity to the temperature that is not affected by the presence of organic phosphates, while the hemoglobin A (HbA) shows more exothermic values of ΔH which decrease when BPG is added.

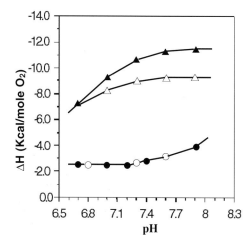

Fig. 3. Effect of temperature on the oxygen affinity of *Dryomys nitedula* (*circles*) and human (*triangles*) hemoglobins in the absence (*filled symbols*) and presence (*open symbols*) of 3 m*M* 2,3-BPG. The values are calculated from van't Hoff equation, using the data obtained from oxygen equilibria experiments and corrected for the heat contribution of oxygen in solution (-3 kcal·mol⁻¹). Conditions: 0.1 *M* Hepes buffer plus 0.1 *M* NaCl

To evaluate this phenomenon at different concentrations of BPG as well, the concentration of organic phosphate was increased in the presence of 0.1 *M* NaCl at 5°, 20°, and 37°C; the change in logP$_{50}$ for *Dryomys nitedula* hemoglobin induced at pH 7.4 can be seen in Fig. 4,. As is evident, the reduction in oxygen affinity at the three temperatures shows a parallel trend: the ΔlogP$_{50}$ calculated between the three temperatures is always equal, both in the absence and in the presence of the different concentrations of BPG.

In Table 2 the values of the equilibrium constants for the binding of BPG to oxy- and deoxy-hemoglobin in the dormouse at 5° and 37°C are reported (in the same table the values for human hemoglobin can also be seen). Thus, the *Dryomys nitedula* hemoglobin is characterized at 37° both in the T and R conformational states by a similar affinity constant for BPG with respect to the corresponding conformational states of HbA. However, at 5°C the situation is completely different: while, in fact, the HbA increases its affinity for BPG, *Dryomys nitedula* Hb does not modify the values.

Table 2. Association (M^{-1}) constants for the binding of BPG with both oxy and deoxy conformation states for human and *Dryomys* hemoglobins at two different temperatures. Conditions: 0.1 *M* Hepes buffer plus 0.1 *M* NaCl pH 7.4

Samples	pK$_{deoxy}$	pK$_{oxy}$
Dryomys nitedula Hb		
5°C	3.30	2.98
37°C	3.30	2.98
Human Hb		
5°C	4.15	3.56
37°C	3.35	2.94

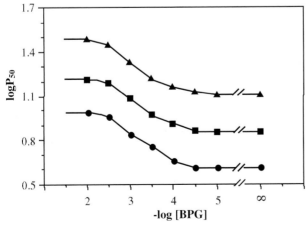

Fig. 4. Effect of 2,3-BPG concentration on the oxygen affinity *Dryomys nitedula* hemoglobin. At 5° (*circles*), 20° (*squares*) and 37°C (*triangles*). Conditions: 0.1 *M* Hepes buffer plus NaCl 0.1 *M* pH 7.4

Discussion

The results obtained for *Dryomys nitedula* hemoglobin show that the effect of Bohr protons, chloride ions, organic phosphates, and temperature is significantly lower than that exerted on human hemoglobins. Thus, the increase in hemoglobin oxygen affinity, resulting from the reduction in red cell organic phosphates and body temperature that occurs during hibernation, is advantageous for loading oxygen in the lungs under conditions of hypoxia, without depressing aerobic tissue metabolism. Furthermore, it is possible that the reduced Bohr effect may moderate the potential effects of increased CO_2 associated with prolonged apnea on the loading and unloading of oxygen.

The overall heat of oxygenation (ΔH) proved to be much less exothermic than that of human hemoglobins and completely independent of BPG concentration. This property may be considered an advantage to maintain a sufficient pressure gradient for oxygen flux between capillary blood and tissues during hibernation, independently of the fluctuations in BPG concentration that dormouse presents in the red cells. In fact, a hemoglobin molecule with these characteristics may represent a compromise which allows the animal, in the hypothermic state, to reap the benefits of enhanced oxygen loading without the risk of oxygen unloading. Furthermore, the low sensitivity to temperature of this hemoglobin limits the sudden variations in oxygen pressure that could occur during the awakening phase when the body temperature returns to 37°C within a few minutes.

In conclusion, the functional and thermodynamic properties observed for dormouse hemoglobin seem to have a physiological significance in relation to the sudden variations in body temperature that this animal encounters during the winter.

References

1. di Prisco G, Condò SG, Tamburrini M, Giardina B (1991) Oxygen transport in extreme enviroments. Trends Biochem Sci 16: 471-474
2. Clementi ME, Condò SG, Castagnola M, Giardina B (1994) Hemoglobin function under extreme life conditions. Eur J Biochem 223: 309-317
3. Kramm C, Sattrup G, Baumann R, Bartels H (1975) Respiratory function of blood in hibernating and non-hibernating hedgehogs. Respir Physiol 25: 311-318
4. Tempel GE, Musacchia XJ (1975) Erythrocyte 2,3-diphosphoglycerate concentrations in hibernating, hypothermic and rewarming hamsters. Proc Soc Exp Biol Med 148: 588-592
5. Doherty JC, Kronon MT, Rotermund AJ (1993) The effects of short term cold storage upon ATP and 2,3-BPG levels in the blood of euthermic and hibernating thirteen-lined ground squirrels spermophilus tridecemlineatus. Comp Biochem Physiol 104A: 87-91
6. Giardina B Amiconi G (1981) Measurement of binding of gaseous and nongaseous ligands to hemoglobin by conventional spectrophotometric procedures. Methods Enzymol 76: 417-427

The Hemoglobin Polymorphism in Sardinian Goats: Nucleotide Sequence and Frequency of β^A, β^D, $\beta^{D\text{-Malta}}$, and β^E Globin Genes*

M. Pirastru[1], M. Palici di Suni[1], G.M. Vacca[2], P. Franceschi[3], B. Masala[1], L. Manca[1]

Introduction

As in most vertebrates, the goat α-globin gene *locus* is duplicated though producing $^I\alpha$ and $^{II}\alpha$ globin chains that differ in three amino acid residues [1-3]. As the result of an unusual evolutionary history, however, a goat β-globin gene cluster consists of 12 genes organized as a triplicated four-gene set located on chromosome 7: 5'-ϵ^I-ϵ^{II}-$\psi\beta^X$-β^C-ϵ^{III}-ϵ^{IV}-$\psi\beta^Y$-β^A-ϵ^V-ϵ^{VI}-$\psi\beta^Z$-β^F-3' [4-6]. The β^A-globin gene is expressed in adult animals, whereas β^F and β^C genes are expressed in the fetus and in juveniles up to 6 months of age, respectively [7]. During the first year of postnatal life, the synthesis of pre-adult β^C-globin is supplanted by the synthesis of adult β^A. The $\beta^C\rightarrow\beta^A$ switch is reversible and the reactivation of the β^C synthesis, at the expense of β^A, can be induced by anemia, hypoxia, or the administration of erythropoietin [1,8]. Due to this unusual feature, which is also common in sheep hemoglobin (Hb) type A and in Sardinian mouflon [5, 7, 9, 10], the organization of globin genes, the nucleotide sequence, and the evolution of goat and sheep globin genes have been extensively studied [1-2, 11-13].

The occurrence of several variants at the two α-*loci* and at the adult β^A-*locus* of goat has been described previously [14, 15]. Three allelic β chains (named β^E, β^D, $\beta^{D\text{-Malta}}$) were detected by conventional electrophoretic and chromatographic techniques and structurally characterized at the protein level [14-19]. The

[1]Dept. of Phisiological, Biochemical and Cell Sciences, University of Sassari, Sassari, Italy
[2]Dept. of Animal Biology, University of Sassari, Sassari, Italy
[3]CEVAREN, University of Corte, Corte, France

The authors wish to dedicate this article to the memory of Professor Titus H. J. Huisman, generous teacher, and invaluable scientist.

G. di Prisco, B. Giardina, R.E. Weber (Eds)
Hemoglobin Function in Vertebrates.
Molecular Adaptation in Extreme and Temperate Environments
© Springer-Verlag Italia 2000

frequency of these allelic chains varies widely in goats from different geographic areas [20, 21]. With the advent of the highly sensitive isoelectric focusing in ultra narrow immobilized pH gradient technique, it soon became evident that the β^A chain may be a mixture of four different allelic chains, termed β^{A2}, β^{A4}, β^{A6}, and β^{A8} [19, 20, 22]. None of these alleles have been characterized with regard to their amino acid or nucleotide sequence so far.

A study was conducted in hundreds of goats living in Sardinia by means of isoelectric focusing (IEF) of Hb tetramers and electrophoresis of dissociated globin chains. This screening showed that Sardinian breeds are also characterized by a notably high Hb polymorphism. Here, we report the results of this screening and our findings on nucleotide sequence analysis of the four most common allelic β-globin genes, all characterized by distinct electrophoretic patterns. These are the β^A, β^D, $\beta^{D-Malta}$, and β^E genes. It is expected that with the methodologies described here the various α and β chain variants present in goat can be completely characterized.

Materials and Methods

Detection of Hb Polymorphism

EDTA blood samples were collected from 340 adult goats living on the island of Sardinia, washed three times with isotonic saline, lysed with CCl_4, and diluted with 0.05% KCN to a final concentration of ca. 1 g Hb/dl. To determine the Hb phenotype, lysates were analyzed by IEF on thin-layer 5% polyacrylamide slab gels in the 6.7-7.7 pH range, as previously described [23,24]. Constituent globin chains were analyzed by the dissociating polyacrylamide slab gels in the presence of acetic acid, 8 M urea, and Triton X-100 [25].

Sequencing of β-Globin Genes

DNA was obtained from white cells of the peripheral blood of animals with the selected Hb phenotypes by a standard method. It has been shown that goat β^A, β^C- and β^F-globin genes share very similar nucleotide sequences [2], so a strategy for the selective amplification of the β^A gene by specific oligonucleotide primers was required. A 1305-bp fragment, containing exon 1, IVS I, exon 2, and most of IVS II of the β^A gene, was obtained using an upstream primer 5'-GCT-GCTGCTTACACTTGCTT-3' (DSB7) that anneals to three β-globin genes at positions -10 to +10, with respect to the Cap site of the genes, and a specific downstream primer 5'-GTCCATGGATTTCTCCAGGC-3' (SB10) that only matches with the β^A gene in a DNA region which is deleted in both β^C and β^F genes (at positions +1276 to +1295). A second pair of primers was used to amplify exon 3. Of the two, SB5 (5'-GTATTCTTGTGCTTCCCTTGTGG-3') was located at positions +1147 to +1168 and was specific for β^A and β^C genes, whereas SB6 (5'-TTCTTTATATCTTCAGTGCTTTGC-3', reverse) was located at positions +1703 to +1726 and was nonspecific. Two different fragments (580 bp and 519 bp in length) containing part of IVS II, exon 3, and part of the 3' UTR of the β^A and β^C genes, respectively, were obtained. The 1305- and 580-bp fragments were isolated and sequenced by the dideoxy chain terminating method [26] of single-stranded DNA using ^{33}P dideoxy and labeled nucleotides and the following primers: DSB7, SB10, SB3 (5'-TGACTTCCTCTGACCTTGT-3', from +230 to +248), and GB4 (5'-CTCGTTCTTTTTATGGTCAA-3', from +518 to +537) to sequence exons 1 and 2, and 5'-GCCTGGAGAAATCCATGGAC-3' (SB11, reverse, from +1276 to +1295), and 5'-GGAAATCAGGAAGGGGAGC-3' (GB2, reverse, from +1525 to +1543) to sequence exon 3.

Results

Hemoglobin Polymorphism

Schon et al. [27] described the amino acid sequence of the goat $^I\alpha$ and $^{II}\alpha$ globin chains as deduced by the nucleotide sequence of the proper coding genes. The two globins, expressed in the approximate 3:1 ratio, differ at positions 19, 113, and 115: $^I\alpha$ chain has Gly, Leu, and Asn, whereas the $^{II}\alpha$ chain has Ser, His, and Ser, respectively. The presence of His instead of Leu at 113 of the $^{II}\alpha$ chain is responsible for the cathodal focalization and resolution of the corresponding tetramer. An allelic $^I\alpha$ chain has been described as containing the Asp→Tyr substitution at position 75 [28]. The Asp→Tyr substitution in this (not very common) allelic chain is responsible for the more anodal than normal focalization of the corresponding Hb. It appears that a possibly as yet undescribed $^{II}\alpha^S$ chain, which focuses less anodally with respect to the normal chain, has been identified in this work. Thus, allelic forms of both $^I\alpha$ and $^{II}\alpha$ chains were observed in this screening.

Figure 1 offers an example of the Hb polymorphism we found by the IEF method. This is due to the presence of the two different nonallelic $^I\alpha$ and $^{II}\alpha$ chains, of the allelic variants at one of the two α-loci and of several variants at the β locus. Hb tetramers containing a single β-globin chain (as those in lanes 1, 2, and 4) are seen as two clearly separated bands: the more anodic, which contains the major $^I\alpha$ chain, and the less anodic, which contains the minor $^{II}\alpha$ chain. The samples in lanes 3 and 7 show an anodic doublet of Hbs due to tetramers containing the $^I\alpha$ and $^{II}\alpha$ chains (i.e., $^I\alpha_2\beta_2$, and $^{II}\alpha_2\beta_2$) and a cathodic doublet due to tetramers containing the same α chains and a different β chain (i.e., $^I\alpha_2\beta^x{}_2$, and $^{II}\alpha_2\beta^x{}_2$). The sample in lane 6 indicates the presence of three different α chains (a major $^I\alpha$ and two allelic $^{II}\alpha$ chains) as the counterpart of a single β chain. The sample in lane 8 has $^I\alpha$ and $^{II}\alpha$ chains in combination with two allelic β chains, whereas the samples in lanes 5 and 9 have three α chains ($^I\alpha$ and two $^{II}\alpha$) in combination with two different β chains. It is nonetheless clear that with IEF, despite its high resolution power, it is not pos-

+

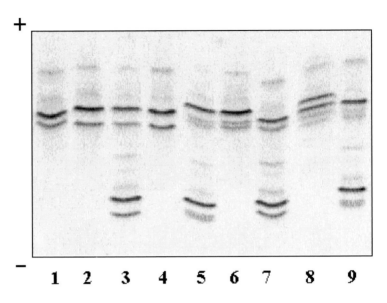

−

1   2   3   4   5   6   7   8   9

Fig. 1. The Hb polymorphism in Sardinian goat as determined by IEF. Several Hb phenotypes in individuals homozygous and heterozygous for different α-globin haplotypes and β-globin alleles are shown. See text for explanations

sible to identify all the α and β normal and variant chains that are present in a given phenotype.

A better understanding of such a polymorphism was possible when each sample was also analyzed in terms of constituent globin chains by the acid-urea gel electrophoresis (AUT-PAGE) method. Figure 2 shows that as many as seven different β chains and four different α chains may coexist in the sample of 340 goats. These different β chains were provisionally indicated as β^1, β^2, β^3, β^X, β^{IX}, β^{2X}, and β^Y, respectively. Both $^I\alpha$ and $^{II}\alpha$ genes display allelic and nonallelic variants. The most common are indicated here as $^I\alpha$ and $^{II}\alpha$, whereas the less common variants are referred to as the $^I\alpha^{Fast}$ (most probably corresponding to the 75Tyr-containing $^I\alpha$ chain) and $^{II}\alpha^{Slow}$ because of their mobility toward the cathode in the IEF and AUT-PAGE systems. Three different tandem associations (α haplotypes) of α chains were observed: $^I\alpha$-$^{II}\alpha$, $^I\alpha$-$^{II}\alpha^S$, and $^I\alpha^F$-$^{II}\alpha$, either in homozygosity or in heterozygosity. The sample in lane 11 is a $^I\alpha$-$^{II}\alpha^S$/$^I\alpha^F$-$^{II}\alpha$ heterozygote, thus showing four different α chains (see legend for other haplotypes). The $^I\alpha$ chain appears to be coded by a gene that was present in linkage with the $^{II}\alpha$ chain-producing gene in 506 out of 680 chromosomes and, in linkage with the $^{II}\alpha^S$ gene, in 162 chromosomes. The "fast-moving" $^I\alpha^F$ chain appears to be coded by a gene, observed in 12 chromosomes, in linkage with the $^{II}\alpha$ gene. Table 1 lists the observed combinations of α haplotypes and allelic β chains, together with the frequency of the three α haplotypes and of the seven β genes. We observed 19 out of the possible 29 combinations of β chains

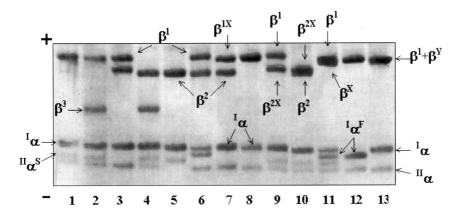

Fig. 2. Separation of constituent, dissociated, globin chains by means of the acid-urea gel electrophoresis (AUT-PAGE). Four different α and seven different β chains are identified. *Lanes 3, 7, 8,* and *13,* $^I\alpha-^{II}\alpha/^I\alpha-^{II}\alpha$; *lane 5,* $^I\alpha-^{II}\alpha^S/^I\alpha-^{II}\alpha^S$; *lane 12,* $^I\alpha^F-^{II}\alpha/^I\alpha^F-^{II}\alpha$; *lanes 1, 2, 4, 9,* and *10,* $^I\alpha-^{II}\alpha^S/^I\alpha-^{II}\alpha$; *lane 6,* $^I\alpha-^{II}\alpha/^I\alpha^F-^{II}\alpha$

Table 1. Hemoglobin phenotypes observed in Sardinian goat

β-alleles / α-haplotype	β^A	β^E	β^D	β^X	$\beta^{D\text{-}MALTA}$	$\beta^A\beta^E$	$\beta^A\beta^D$	$\beta^A\beta^X$	$\beta^A\beta^{D\text{-}MALTA}$	$\beta^A\beta^{2X}$	$\beta^A\beta^Y$	$\beta^E\beta^D$	$\beta^E\beta^X$	$\beta^E\beta^{D\text{-}MALTA}$	$\beta^E\beta^{2X}$	$\beta^E\beta^X$	$\beta^D\beta^{D\text{-}MALTA}$	$\beta^D\beta^{2X}$	$\beta^X\beta^{D\text{-}MALTA}$	Total
$^I\alpha-^{II}\alpha/^I\alpha-^{II}\alpha$	55	8	7	1	1	35	39	1	13	5	1	15	2	5	?	1	3	-	-	193
$^I\alpha-^{II}\alpha^S/^I\alpha-^{II}\alpha^S$	8	2	-	-	-	3	6	-	1	-	-	3	-	1	-	-	1	-	-	25
$^I\alpha^F-^{II}\alpha/^I\alpha^F-^{II}\alpha$	1	-	-	-	-	-	-	-	-	-	-	-	-	-	-	-	-	-	-	1
$^I\alpha-^{II}\alpha/^I\alpha-^{II}\alpha^S$	35	-	4	-	-	16	27	2	9	3	-	7	-	2	2	-	3	1	-	111
$^I\alpha-^{II}\alpha/^I\alpha^F-^{II}\alpha$	2	-	-	-	-	2	2	-	1	-	-	-	-	1	-	-	-	-	1	9
$^I\alpha-^{II}\alpha^S/^I\alpha^F-^{II}\alpha$	-	-	-	-	-	-	-	1	-	-	-	-	-	-	-	-	-	-	-	1
Total	101	10	11	1	1	56	74	4	24	8	1	25	2	9	3	1	7	1	1	340

and all of the six possible combinations of α haplotypes. Table 1 shows that as many as 45 of the theoretically possible 168 different Hb phenotypes, which resulted from the combinations of the above allelic and nonallelic variants, were observed. We assumed that the β^1 gene, the most frequent allele in our sample ($f=0.543$), would be the β^A gene previously described [1, 2].

DNA Sequencing Data

Genomic DNA was sequenced from animals with the globin chains we indicated as $\beta^1, \beta^2, \beta^3$ and β^{1X}, which were the most common variants in the sample. The nucleotide sequence and protein sequence encoded by the three coding blocks of the β^1 gene agree completely with the wild β^A allele described by Schon et al. [2] and with the amino acid sequence of the β^A globin previously reported [14, 16]. This sequence was thus considered as the reference sequence in order to identify one of the variant chains already described and, possibly, new variants. Figure 3a shows part of the sequencing gels in the region of codons 19 to 23 of a goat homozygous for the β^1 gene, a goat heterozygous for the β^1 and the β^3 genes, and one homozygous for the β^3 gene. Homozygosity is evident at position 21 for the $\underline{G}$AT codon (Asp) in the β^1/β^1 goat, and for the $\underline{C}$AT codon (His) in the β^3/β^3 goat. Heterozygosity for the two changes at the same codon is evident in the β^1/β^3 goat (base changes are depicted in underlined bold). Similarly, Fig. 3b shows the sequencing in the region of codons 67 to 71 in a β^1/β^1 and in a β^1/β^{1X} animal, indicating that G$\underline{A}$C (Asp) is in β^1, whereas G$\underline{G}$C (Gly) is in β^{1X} at position 69. In Fig. 4 the complete sequence of β^A and that of other sequenced genes is aligned. The amino acid substitutions observed correspond to three common variants already described at the protein level: β^E (the β^2), β^D (the β^3), and $\beta^{D\text{-Malta}}$ (the β^{1X}) genes [15-18]. Sequence comparison at the coding region level shows that β^D and $\beta^{D\text{-Malta}}$ genes only have one substitution due to a single base change ($\underline{G}$AT→$\underline{C}$AT at position 21, corresponding to the Asp→His substitution, and G$\underline{A}$C→G$\underline{G}$C at position 69, corresponding to the Asp→Gly substitution, respectively). The β^E gene has five substitutions: three were consistent with transitions at codons 78, 95, and 104 and two were transvertions at codons 87 and 125. Substitutions at codons 78 and 95 result in synonymous codons (CT$\underline{T}$→CT$\underline{C}$ for Leu, AA$\underline{G}$→AA$\underline{A}$ for Lys, respectively), while substitutions at codons 87, 104, and 125, being missense mutations, correspond to the Gln→His, Lys→Arg and Leu→Val changes. At the level of intron sequences, the β^E gene shows the higher number of substitutions (six).

Discussion

In this report we showed that a notable Hb polymorphism characterizes Sardinian goats. The four most common alleles at the β *locus* have the following gene frequencies: β^A=0.543, β^E=0.169, β^D=0.191, $\beta^{D\text{-Malta}}$=0.063. The remaining three alleles, the structures of which are under investigation, have lower frequencies, ranging from 0.015 to 0.0015. Work conducted over three decades has identified the Hb polymorphism in most breeds of goats [15,29]. Recent data show that the β^A gene was almost exclusive in the Nubian breeds (f=0.94), frequently observed in the Angora (f=0.857), and less frequently in the Spanish (f=0.517) breeds [21]. The β^E gene had a relatively high occurrence in the Alpine, Saanen, and Spanish breeds (f=0.68, 0.546, and 0.4, respectively). The β^D allele was not found in the Alpine and Angora breeds. Extensive variation of goat Hb was discovered by application of the very high resolution Immobiline electrophoresis technique [19,20,30]. By means of this method, the Hb A was further separated into four variants (A2, A4, A6, and A8) in animals of Norwegian breeds. In a (limited) sample of Norwegian Saanen goats, the frequency of the β^A gene (A4+A6+A8) was 0.75, that of β^E was 0.03, and of β^D 0.22 [30], whereas the α haplotype frequency was very similar to that observed in this work ($^I\alpha$-$^{II}\alpha$=0.744, $^I\alpha$-$^{II}\alpha^S$=0.238, respectively). Manwell and Baker [31] suggested that man has played an important role in generating polymorphisms in domesticated mammals by breeding from individual that would otherwise

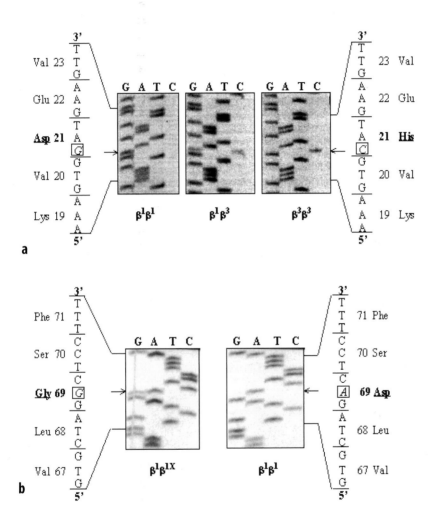

Fig. 3a,b. a The nucleotide sequencing of part of the β-globin gene, exon 1, showing the presence of the $\underline{G}$AT codon (Asp) at position 21 in an individual homozygous for the β^1 allele, the presence of the $\underline{C}$AT sequence (His) at the same position in an individual homozygous for the β^3 allele, and of both sequences in a heterozygous goat. **b** The nucleotide sequencing of part of the exon 2, codons 67 to 71, in a β^1/β^1 goat and in a β^1/β^{1X} goat, showing that G$\underline{A}$C (Asp) is in β^1, whereas G$\underline{G}$C (Gly) is in β^{1X} at position 69. The β^1 gene was identified as the β^A wild gene whereas the β^{1X} was identified as the $\beta^{D\text{-Malta}}$ allele

```
                                              2                    10
              +25                   MetLeuThrAlaGluGluLysAlaAlaAlaValThrGlyPheTrpGlyLysValL
βᴬ            ttcactagcagctacacaaacagacaccATGCTGACTGCTGAGGAGAAGGCTGCCGTCACCGGCTTCTGGGGCAAGGTGA 80
βᴰ⁻ᴹᵃˡᵗᵃ      ................................................................................
βᴰ            ................................................................................
βᴱ            ................................................................................

                   20                      30
              ysValAspGluValGlyAlaGluAlaLeuGlyAr
βᴬ            AAGTGGATGAAGTTGGTGCTGAGGCCCTGGGCAGgtaggtatctcacttacaagacaggtttaaggagagtggatggcac 16●
βᴰ⁻ᴹᵃˡᵗᵃ      ................................................................................
βᴰ            .....C..........................................................................
βᴱ            ................................................................................


βᴬ            ctaggcatgtgaggacagagccgtccctgagattctgaaagctgctgacttcctctgaccatgtgctgtttcatccctt 24●
βᴰ⁻ᴹᵃˡᵗᵃ      ................................................................................
βᴰ            ................................................................................
βᴱ            ...........:....................................................................

                                         40                            50
              gLeuLeuValValTyrProTrpThrGlnArgPhePheGluHisPheGlyAspLeuSerSerAlaAspAlaValMetAs
βᴬ            agGCTGCTGGTTGTCTACCCCTGGACTCAGAGAGGTTCTTTGAGCACTTTGGGGACTTGTCCTCTGCTGATGCTGTTATGAA 32●
βᴰ⁻ᴹᵃˡᵗᵃ      ................................................................................
βᴰ            ................................................................................
βᴱ            ................................................................................
                              60                            70                            80
              nAsnAlaLysValLysAlaHisGlyLysLysValLeuAspSerPheSerAsnGlyMetLysHisLeuAspAspLeuLysG
βᴬ            CAATGCTAAGGTGAAGGCCCATGGCAAGAAGGTGCTAGACTCCTTTAGTAACGGCATGAAGCATCTTGACGACCTCAAGG 40●
βᴰ⁻ᴹᵃˡᵗᵃ      .................................G..............................................
βᴰ            ................................................................................
βᴱ            ...........................................................................C....
                              90                           100
              lyThrPheAlaGlnLeuSerGluLeuHisCysAspLysLeuHisValAspProGluAsnPheLys
βᴬ            GCACCTTTGCTCAGCTGAGTGAGCTGCACTGTGATAAGCTGCACGTGGATCCTGAGAACTTCAAGgtgagtttgtggagt 48●
βᴰ⁻ᴹᵃˡᵗᵃ      ................................................................................
βᴰ            ................................................................................
βᴱ            ...........C...................A..................................G..............


βᴬ            cctcagtgttttcctcgttcttttttatggtcaagctcattttgtggggagaaggctgaatgacaggacacagtttagaat 56●
βᴰ⁻ᴹᵃˡᵗᵃ      ................................................................................
βᴰ            ................................................................................
βᴱ            ...............................................a..................g..a...........


βᴬ            ggagaagaggtattctggttagagtgctaaggactcctcagaaccgtttagactcttttaacctctttgctcacaaccat 64●
βᴰ⁻ᴹᵃˡᵗᵃ      ................................................................................
βᴰ            ................................................................................
βᴱ            ................................................................................


βᴬ            catttcctctgattcattcttgttctctgttgtctgcaatgtcctcttttttaattatacttttttattttgagggtttaat 72●
βᴰ⁻ᴹᵃˡᵗᵃ      ................................................................................
βᴰ            ................................................................................
βᴱ            ................................................................................


βᴬ            ttgaaaaaaatatttattttatcaactttaaaaatcgtatctagtattttccccttatctgtttctttcaaggaataaat 80●
βᴰ⁻ᴹᵃˡᵗᵃ      ................................................................................
βᴰ            ................................................................................
βᴱ            ................................................................................
```

```
βᴬ        gttctattgcttttttgaaatgattcaaaataataaaaatgataacaagttctggattaagttagaaagagagaaacattt 880
βᴰ⁻ᴹᵃˡᵗᵃ   ................................................................................
βᴰ        ................................................................................
βᴱ        ................................................................................

βᴬ        ctaaatatatattcaggaagagatataggtagatacatatcagtagtaacatcttcacttcagtcatccttgtgcttatatc 960
βᴰ⁻ᴹᵃˡᵗᵃ   ................................................................................
βᴰ        ................................................................................
βᴱ        ................................................................................

βᴬ        ctatggtcacagcttgggatgagactgaaataccctgaatctaaccttggacttctctcatagctcggttggtaaagagt 1040
βᴰ⁻ᴹᵃˡᵗᵃ   .......................................................................a...........
βᴰ        .......................................................................a...........
βᴱ        ...c...................................................................a...........

βᴬ        ctgcctgcaatgcaggagatcccagttcgattcctgggtcaggaaaagaatggctggagaagggataggctacccactcc 1120
βᴰ⁻ᴹᵃˡᵗᵃ   ................................................................................
βᴰ        ................................................................................
βᴱ        ................................................................................

βᴬ        agtattcttgtgcttcccttgtggctcagctggtaaagaatctgcctgcagtgcgggagacctgggttcttctatccatg 1200
βᴰ⁻ᴹᵃˡᵗᵃ   ................................................................................
βᴰ        ................................................................................
βᴱ        ................................................................................

βᴬ        ggttgggaagatccctggagaagggaaaggctaccctctccagtattctggcctggagaaatccatggactgtatagtc 1280
βᴰ⁻ᴹᵃˡᵗᵃ   ................................................................................
βᴰ        ................................................................................
βᴱ        ................................................................................

βᴬ        catagggttgcaaagagtcagacatgactgagcaactttcactttactaacctgcactaaccctgcccttgcttaatgtc 1360
βᴰ⁻ᴹᵃˡᵗᵃ   ................................................................................
βᴰ        ................................................................................
βᴱ        ................................................................................
                              110                         120
          LeuLeuGlyAsnValLeuValValValLeuAlaArgHisHisGlySerGluPheThrProLeuLeuGln
βᴬ        tttccacacagCTCCTGGGCAACGTGCTGGTGGTTGTGCTGGCTCGCCACCATGGCAGTGAATTCACCCCGCTGCTGCAG 1440
βᴰ⁻ᴹᵃˡᵗᵃ   ................................................................................
βᴰ        ................................................................................
βᴱ        ...........................................................................G........
                  130                         140
          AlaGluPheGlnLysValValAlaGlyValAlaAsnAlaLeuAlaHisArgTyrHisStop
βᴬ        GCTGAGTTTCAGAAGGTGGTGGCTGGTGTTGCCAATGCCCTGGCCCACAGATATCACTAAgctccccttcctgatttcca 1520
βᴰ⁻ᴹᵃˡᵗᵃ   ................................................................................
βᴰ        ................................................................................
βᴱ        ................................................................................

βᴬ        ggaaaggtttttctgttctcaaagaccaaaaattgaatatggaaaaattatgaagcatttgagcacctggcctctgctt 1600
βᴰ⁻ᴹᵃˡᵗᵃ   ................................................................................
βᴰ        ................................................................................
βᴱ        ............a...................................................................
                    +1640
βᴬ        aataaagacacttatt
βᴰ⁻ᴹᵃˡᵗᵃ   ................
βᴰ        ................
βᴱ        ................
```

Fig. 4. Alignment of the complete sequences of the βᴬ, βᴱ, βᴰ, and βᴰ⁻ᴹᵃˡᵗᵃ globin genes. Sequences are from position +25 to +1640 with respect to the Cap site. The coding sequences, in *capital letters*, are indicated with the three-letter amino acid code. The polyadenylation signal is *underlined*

be geographically isolated. Thus, for a better understanding of Hb polymorphism in goats, the sample to be analyzed should cover as wide a range as possible. In this respect, our data on allele occurrence and frequency offer two advantages: the range of the sample is wide enough and the sample was collected in the island of Sardinia, a land where over the past 6000 years sheep are the domesticated mammals which were certainly inbred most , whereas goats received little attention.

Although the long evolutionary history of the goat β-globin cluster may be an explanation, such a high degree of Hb polymorphism is intriguing but still not completely understood. The reason for the high rate of polymorphism in the goat β-globin *locus* is not known. According to the sequence of the β^D and $\beta^{D-Malta}$ alleles, we might assume that these could have originated by the wild β^A allele following independent single-point mutation events, which occurred at codons 21 and 69, respectively. On the other hand, the β^E allele is the most interesting, having three substitutions at positions 87 (Gln→His), 104 (Lys→Arg), and 125 (Leu→Val). In the 3' block of sheep β^A chain, Arg is at 104 and Val is at 125, so that it may be assumed that the β^E gene is the product of a recombination event between goat and sheep β^A genes which occurred at the time of goat/sheep divergence [32]. The synonymous CT*T*→CT*C* for Leu and AA*G*→AA*A* for Lys substitutions at codons 78 and 95, and the mutation at 87, might have occurred after the β^E gene arose. It is expected that determining the remaining three β^X, β^{2X}, and β^Y alleles might contribute to the understanding of goat Hb polymorphism.

Acknowledgements

This work was supported in part by INTERREG II and Structural Funds of the European Community and refers parts of the material described by M. Pirastru in her thesis for Dottorato di Ricerca degree of the University of Sassari, Italy.

References

1. Huisman THJ, Adams HR, Dimmock MO, Edwards WE, Wilson JB (1967) The structure of goat hemoglobins. J Biol Chem 242:2534-2541
2. Schon EA, Cleary ML, Haynes JR, Lingrel JB (1981) Structure and evolution of goat γ-, β^c- and β^A-globin genes: three developmentally regulated genes contain inserted elements. Cell 27:359-369
3. Kleinschmidt T, Sgouros JG (1987) Hemoglobin sequences. Biol Chem Hoppe-Seyler 368:579-615
4. Townes TM, Fitzgerald MC, Lingrel JB (1984) Triplication of a four-gene set during evolution of the goat β-globin locus has produced three genes now expressed differentially in development. Proc Natl Acad Sci USA 81:6589-6593
5. Garner KJ, Lingrel JB (1989) A comparison of the β^A- and β^B-globin gene clusters of sheep. J Mol Evol 28:175-184
6. Simi LB, Sasi R, Lingrel JB, Lin CC (1989) Mapping of the goat β-globin gene cluster to a region of chromosome 7 by in situ hybridization. J Hered 80:246-249

7. Huisman THJ, Lewis JP, Blunt MH, Adams HR, Edwards WE, Wilson JB (1969) Hemoglobin C in newborn sheep and goats: a possible explanation for its function and biosynthesis. Pediatr Res 3:189-198

8. Boyer SH, Hathaway P, Pascasio F, Orton C, Bordley J, Naughton MA (1966) Hemoglobins in sheep: multiple differences in amino acid sequences of three beta-chains and possible origins. Science 153:1539-1543

9. Hammerberg B, Brett I, Kitchen H (1974) Ontogeny of hemoglobins in sheep. Ann NY Acad Sci 241:672-682

10. Masala B, Manca L, Cocco E, Ledda S, Naitana S (1991) Kinetics of the ontogenic and reversible hemoglobin switching in the mouflon (Ovis musimon) and sheep x mouflon hybrid. Comp Biochem Physiol 100A:675-680

11. Di Gregorio P, Rando A, Masina P (1987) Differences in the DNA restriction patterns between sheep with Hb A and Hb B. Anim Genet 18:241-247

12. Rando A, Di Gregorio P, Masina P (1989) Differences in the number of embryonic and pseudo-β-globin genes between Hb A and Hb B sheep. Biochem Genet 27:91-98

13. Rando A, Di Gregorio P, Capuano M, Senese C, Manca L, Naitana S, Masala B (1996) A comparison between the β-globin gene clusters of domestic sheep (Ovis aries) and Sardinian mouflon (Ovis gmelini musimon). Genet Sel Evol 28:217-222

14. Huisman THJ, Wilson JB, Adams HR (1967) The heterogeneity of goat hemoglobin: Evidence for the existence of two nonallelic and one allelic α chain structural genes. Arch Biochem Biophys 121:528-530

15. Huisman THJ (1970) Multiple α and β chain structural genes as a basis for hemoglobin heterogeneity of the adult goat. In: Peeter H (ed) Protides of the biological fluids, Proceedings, 17th Colloquium Bruges. Pergamon, Oxford, pp 242-248

16. Adams HR, Boyd EM, Wilson JB, Miller A, Huisman THJ (1968) The structure of goat hemoglobins. III. Hemoglobin D, a β chain variant with one apparent amino acid substitution. Arch Biochem Biophys 127:398-405

17. Wrightstone RN, Wilson JB, Miller A, Huisman THJ (1970)The structure of goat hemoglobins. IV. A third β chain variant (β^E) with three apparent amino acid substitutions. Arch Biochem Biophys 138:451-456

18. Bannister JV, Bannister WH, Wilson JB, Lam H, Miller A, Huisman THJ (1979) The structure of goat haemoglobins. V. A fourth β chain variant (β-D-Malta; 69Asp→Gly) with decreased oxygen affinity and occurring at a high frequency in Malta. Hemoglobin 3:57-75

19. Braend M, Tucker EM, Clarke SW (1987) Search for genetic variation in the blood of Norwegian dairy goats reveals a new polymorphism at the Hb β^A chain locus. Anim Genet 18:75-79

20. Braend M, Nesse LL, Efremov GD (1987) Expression and genetics of caprine haemoglobins. Anim Genet 18:223-231

21. Wang S, Foote WC, Bunch TD (1990) Transferrin and hemoglobin polymorphism in domesticated goats in the USA. Anim Genet 21:91-94

22. Bergersen O, Braend M (1990) Characterization by RFLP analysis of the caprine β-globin gene cluster in Norwegian dairy goats. Hemoglobin 14: 87-102

23. Naitana S, Ledda S, Cocco E, Manca L, Masala B (1990) Hemoglobin phenotype of the European mouflon sheep living on the Island of Sardinia. Anim Genet 21:67-75

24. Masala B, Manca L (1991) Detection of the common Hb F Sardinia [^Aγ75(E19)Ile→Thr] variant by isoelectric focusing in normal newborn and in adult affected by elevated fetal hemoglobin syndromes. Clin Chim Acta 198:195-202

25. Manca L, Formato M, Demuro P, Pilo G, Gallisai D, Orzalesi M, Masala B (1986) The

gamma globin chain heterogeneity of the Sardinian newborn baby. Hemoglobin 10:519-528

26. Sanger F, Nicklen S, Coulson AR (1977) DNA sequencing with chain-terminating inhibitors. Proc Natl Acad Sci USA 74:5463-5467

27. Schon EA, Wernke SM, Lingrel JB (1982) Gene conversion of two functional goat α-globin genes preserves only minimal flanking sequences. J Biol Chem 257:6825-6835

28. Huisman THJ, Brandt G, Wilson JB (1968) The structure of goat hemoglobins. II. Structural studies of the alpha chains of the hemoglobins A and B. J Biol Chem 243:3675-3686

29. Blunt MH, Huisman THJ (1975) The hemoglobin of sheep. In: Blunt MH (ed) The blood of sheep. Springer, Berlin Heidelberg New York, pp 155-183

30. Braend M, Tucker EM (1988) Hemoglobin types in Saanen goats and Barbary sheep: genetic and comparative aspects. Biochem Genet 26:511-518

31. Manwell C, Backer CMA (1976) Protein polymorphism in domesticated species: evidence for hybrid origin? In: S Karlin, E Nevo (eds) Population genetics and ecology. Academic, New York, pp 105-139

32. Garrick MD, Garrick LM (1983) Hemoglobins and globin genes. In: Agar NS, Board PG (eds) Red blood cells of domestic mammals. Elsevier, Amsterdam, pp 165-207

The Organization of the β-Globin Gene Cluster and the Nucleotide Sequence of the β-Globin Gene of Cyprus Mouflon (*Ovis gmelini ophion*)*

E. Serreri[1], E. Hadjisterkotis[2], S. Naitana[3], A. Rando[4], P. Ferranti[5], M. Corda[6], L. Manca[1], B. Masala[1]

Introduction

The β-globin gene cluster of the domestic sheep *(Ovis aries)* shows two common haplotypes: the A haplotype, which bears the adult β^A allele (HBB^A), and the B haplotype, which bears the adult β^B allele (HBB^B) [1, 2]. The chromosomal organization of the A haplotype was found to be similar to that present in goat *(Capra hircus)* since it shows the same triplication (5'-ϵ^I-ϵ^{II}-$\psi\beta^I$-β^C-ϵ^{III}-ϵ^{IV}-$\psi\beta^{II}$-β^A-ϵ^V-ϵ^{VI}-$\psi\beta^{III}$-β^F-3') of an ancestral four gene set (ϵ-ϵ-$\psi\beta$-β) [3]. This set is characterized by the presence of two embryonic genes (ϵ), one pseudogene ($\psi\beta$), and one adult gene (β). The expression of the adult gene varies during ontogenic development and under different physiological conditions. The β^C, β^A, and β^F genes are, in fact, expressed during juvenile, adult and fetal life, respectively, and the β^C gene expression is reactivable, at the expense of β^A gene, under particular physiological or experimental conditions such as anemia and hypoxia or the administration of erythropoietin [4-6]. The B haplotype is considered to have diverged from the A haplotype, as the result of a recent deletion from a triplicated *locus*. In fact, due to the lack of the whole juvenile four-gene set containing the β^C gene, it is duplicated (5'-ϵ^I-ϵ^{II}-$\psi\beta^I$-β^B-ϵ^{III}-ϵ^{IV}-$\psi\beta^{II}$-β^F-3') and sheep which are homozygous for the β^B allele do not exhibit the $\beta^B \rightarrow \beta^C$ switching.

[1]Dept. of Phisiological, Biochemical and Cell Sciences, University of Sassari, Sassari, Italy
[2]Game and Fauna Service, Ministry of the Interior, Nicosia, Cyprus
[3]Dept. of Animal Biology, University of Sassari, Sassari, Italy
[4]Institute of Animal Production, University of Basilicata, Potenza, Italy
[5]Dept. of Organic and Biological Chemistry, University of Naples, Naples, Italy
[6]Dept. of Sciences Applied to Biosystems, University of Cagliari, Monserrato, Italy

*The authors wish to dedicate this article to the memory of Professor Titus H. J. Huisman, generous teacher, and invaluable scientist.

G. di Prisco, B. Giardina, R.E. Weber (Eds)
Hemoglobin Function in Vertebrates.
Molecular Adaptation in Extreme and Temperate Environments
© Springer-Verlag Italia 2000

Several authors documented the close genetic similarity of wild mouflon (*O. gmelini musimon*) and domestic sheep. Some [7-11] consider that the mouflon were autochthonous forms, which are the ancestors of the present-day domestic sheep. Others [12, 13] claimed that in the Corsico-Sardinian islands mouflon originated by feralization of the first semi-domesticated Neolithic sheep [14]. It is not clear, however, whether this was indeed an ancestral sheep. The literature contains conflicting information on the taxonomy of the Mediterranean mouflon. Earlier authors often classified these insular caprines as specific or subspecific geographic forms, almost entirely on the basis of arbitrary criteria and of the examination of scattered materials [14]. Pfeffer [15] considered the Eurasian sheep, excluding the snow sheep, as one species (*O. ammon*). He considered the mouflon of Corsica and Sardinia "absolument identiques" with the mouflon of Cyprus, under the name *O. a. musimon*. Van Haaften [16] named the Cyprus mouflon as *O. a. orientalis* var. *Zypern* (Cyprus). Schaller [17], considering Pfeffer's work, noted that grouping the small and distinctive urials with the large argalis negates the purpose of taxonomy by obscuring rather than clarifying relationships. Finally, Cugnasse [18] reviewed the taxonomy of the European mouflon and proposed that the following nomenclature be applied to wild sheep living on the Mediterranean islands of Sardinia, Corsica, and Cyprus, respectively: *Ovis gmelini musimon* var. *musimon*, *Ovis gmelini musimon* var. *corsicana*, and *Ovis gmelini musimon* var. *ophion*. He considered the three island populations as being of the same subspecies, with the same origin but a different evolution because of the island isolation. In 1997, the IUCN/SSC Caprinae Specialist Group [19] decided that the Cyprus mouflon is a different subspecies than the European mouflon, named *O. orientalis ophion*. Hadjisterkotis [20], considering the differences between the Cyprus and the European mouflon, noted that the Cypriot wild sheep is a distinctive subspecies under the name *O. g. ophion*. Priority was given to the name *O. gmelini* instead of *O. orientalis* because the first specimen described as *O. orientalis* was a hybrid [20, 21].

New information has recently been published concerning the organization and function of mouflon globin genes and hemoglobins (Hbs), and the way in which they compare with those of sheep. In a hundred wild mouflon captured in Sardinia, two alleles at the adult β-globin *locus* were observed: the HBB^B and the HBB^M, with frequencies of 0.94 and 0.06, respectively [22]. Both β-globin chains can be distinguished from the β^A and β^B chains observed in sheep by electrophoretic techniques. A study at the molecular level [23] indicated that the Sardinian mouflon β^B gene lies on a triplicated β-globin gene cluster homologous to sheep β^A, whereas the β^M allele lies on a duplicated one, homologous to sheep β^B. The mouflon β^B globin chain is, in fact, more similar to sheep β^A (two amino acid differences at positions 75 and 76) than to sheep β^B (five differences at positions 50, 58, 120, 129, and 144) [24]. As a result, mouflon Hb B and sheep Hb A are very similar in their response to organic anions and protons, whereas sheep Hb B displays an oxygen affinity lower than that of mouflon Hb B and sheep Hb A [25]. On the other hand, mouflon Hb M has the same oxygen affinity as the sheep Hb B (unpublished observations). It seems clear, as has been already pointed out [22], that the current nomenclature of sheep and mouflon Hbs, merely based upon the rate of electrophoretic mobility, is inadequate and misleading.

In a preliminary report [26], it has been shown that mouflon living on the island of Cyprus has a "slow-moving" Hb fraction, clearly distinguishable from Sardinian mouflon Hb B by isoelectric focusing (IEF), which is not associated with the production of Hb C during anemia, thus indicating the β-globin gene of this species is on a duplicated haplotype homologous to sheep β[B]. The purpose of this article was to characterize the β-globin chain and gene of Cyprus mouflon and to assess the organization of the β-globin gene cluster. An additional aim was to contribute to the understanding of the origin and evolution of wild mouflon and domestic sheep and to the different ways they adapted to the environment. Data show that the Cyprus mouflon β-globin gene and sheep β[B] gene have a high degree of homology, differing in only one base change at the coding region, leading to a single amino acid substitution (Lys→Arg at codon 144), and in 0.25 % substitutions in the noncoding regions. This single amino acid change, however, might be responsible for molecular adaptation to different environmental conditions.

Materials and Methods

Blood Samples

Blood samples from 18 wild mouflon captured at various locations throughout the Paphos Forest of the island of Cyprus were collected in EDTA as anticoagulant. Washed red cells were lysed by adding 1 volume of cold water and 0.5 volume of CCl_4. Membranes were removed by a 20-min centrifugation at 12,000 g. Sardinian mouflon of the Hb B and Hb M phenotypes and domestic sheep of the Hb A and Hb B phenotypes served as controls.

Gel Electrophoresis of Hb Tetramers and Globin Chains

Hb phenotypes were determined by means of IEF on thin layer 5% polyacrylamide slab gels in the presence of 2.5% ampholytes pH 6.7-7.7, and 0.2% ampholytes pH 3.5-10, as previously described [27]. Dissociated globin chains were analyzed by electrophoresis on 12% polyacrylamide gels containing 6 M urea and 2% Triton X-100, in 5% acetic acid acid-urea gel electrophoresis (AUT-PAGE), as previously described [28].

Isolation of the β-Globin Chain and Protein Sequencing

Lysates containing approximately 100 μg Hb were submitted to reversed-phase high performance liquid chromatography (RP-HPLC), as described previously [29]. The β-globin chains were collected and freeze-dried for the subsequent structural characterization. Automated Edman degradation was performed using an Applied Biosystems (Warrington, U.K.) mod. 477A Protein Sequencer with on-line phenylthiohydantoinyl amino acid (PTHaa)-HPLC analyzer (Perkin

Elmer). Tryptic and carboxypeptidase B digestion was performed in 0.4% ammonium bicarbonate, pH 8.5, at 37°C (E:S 1:50 w/w) for 4 h and 10 min, respectively. Endoproteinase Asp-N digestion was carried out in 0.4% ammonium bicarbonate containing 10% acetonitrile, pH 8.5, at 37°C (E:S 1:100 w/w) for 18 h. Endoproteinase Glu-C digestion was performed in 0.4% ammonium bicarbonate, pH 8.0, at 40°C (E:S 1:100 w/w) for 18 h. Manual Edman degradation steps were carried out directly on the peptide mixtures using 5% phenylisothiocyanate in pyridine as coupling agent, as described previously [30]. Typical mapping experiments were carried out by using 100 μg of intact globin; FAB mass spectra were recorded on a VG-Analytical ZAB-ZSE double focusing mass spectrometer equipped with a VG cesium gun operating at 25 KeV (2 μA) at a resolution of 4000; spectra were recorded on UV sensitive paper and manually counted. Samples were dissolved in 0.1 M HCl and loaded onto a glycerol-thioglycerol-coated probe tip. All the mass values recorded in the FAB mode are shown as monoisotopic masses. Electrospray mass spectrometric (ES/MS) analysis for the intact globins was performed with a BIO-Q triple-quadrupole mass spectrometer (VG, Manchester, UK). HPLC-purified globin samples (10 ml, 25-50 pmol) were injected into the ion source at a flow rate of 2 ml/min; the spectra were scanned from m/z 1600 to 600 at 10 s/scan. Mass-scale calibration was carried out by using the multiple charged ions of a separate introduction of myoglobin. All ES/MS molecular masses are reported as average masses.

Southern Blotting

DNA was obtained from white cells [31]. Southern blot analysis was achieved on DNA samples obtained from mouflon and sheep of both the Hb A and Hb B phenotypes according to Rando [32]. DNA samples were digested with *Hind* III, *Bam* HI, and *Eco* RI and probed with plasmid pGl6Ec3Bm2 (containing the 5' of the goat ϵ^{IV}-globin gene) and plasmid pGγ5' (containing the 5' of the goat β^F-globin gene). In the hybridization conditions reported [32], these plasmids strongly cross-hybridize with the paralogous genes.

Polymerase Chain Reaction and Gene Sequencing

A fragment 2109 base pair (bp) long containing the entire β-globin of both mouflon and sheep was obtained by the polymerase chain reaction (PCR) amplification technique using the Sb1 forward primer (5'-AATAATCCATCCA CATAGTCTTGAA-3') located from positions -392 to -368 with respect to the Cap site, and the Sb6 reverse primer (5'-TTCTTTATATCTTCAGTGCTTTGC-3') located from positions +1694 to +1717.

Direct sequencing of the PCR-amplified product was carried out using a cycle sequencing procedure with the dideoxy terminator method [33]. Dideoxy terminators were labeled with ^{33}P and the sequencing products were visualized by autoradiography. Automated fluorescent DNA sequencing was also performed (ALF Sequencer, Amersham-Pharmacia Biotech). The 2109-bp fragment was sequenced using Sb1 and Sb6 and the following internal primers: DSb7 (5'-GCTGCCGCTTACACTTGCTT-3', from -10 to +10); Sb3 (5'-TGACTTCCTCT-GACCTTGT-3', from +229 to +247); Sb10 (5'-GTCCATGGATTCTCCAGGC-3', reverse, from +1266 to +1285).

Results

Figure 1 shows the IEF and AUT-PAGE patterns of Hbs and globins from mouflon and sheep. Separation by IEF in Fig. 1a indicates that Hb of Cyprus mouflon has a mobility which is fairly different than sheep Hb B and Sardinian mouflon Hb M. Separation of globin chains (Fig. 1b) more clearly shows that the constituent β-chain of Cyprus mouflon (hereafter termed the β^{Cyprus} chain) moves cathodally with respect to the β^{B}- and the β^{M}-chain of Sardinian mouflon, whereas Fig. 1c clearly outlines the differences in mobility of the sheep β^{A} and β^{B}, and β^{Cyprus} globin chains. On the whole, the result presented in Fig. 1 denotes the β^{Cyprus} globin as being a chain with a polypeptide sequence different from both mouflon and sheep β-chains described so far.

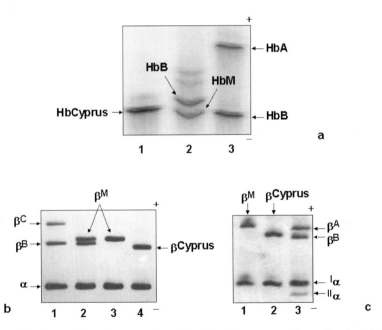

Fig. 1a-c. The electrophoretic properties of the Hb of Cyprus mouflon. **a** Isoelectric focusing of native tetramers. *1*, Cyprus mouflon; *2*, Sardinian mouflon of the Hb B/M phenotype; *3*, sheep of the Hb A/B phenotype. **b** Acid-urea gel electrophoresis (AUT-PAGE) of dissociated globin chains. *1*, Anemic Sardinian mouflon of the Hb B type showing the re-synthesis of the β^{C} chain; *2*, Sardinian mouflon of the Hb B/M type; *3*, Sardinian mouflon of the Hb M type; *4*, Cyprus mouflon. **c** AUT-PAGE of dissociated globin chains showing the position of the β^{B}-Cyprus as compared with the mouflon β^{M} and sheep β^{A} and β^{B} chains

Analysis of tryptic digestion of the Cyprus mouflon β-globin chain by FAB/MS demonstrated that one amino acid replacement was present with respect to ovine β^B sequence: the signal at m/z 1149, corresponding to peptide 133-144, was shifted at m/z 1177. The mass difference between these two signals (Δm=+28) was accounted for by the amino acid replacement of the Lys residue at position 144 with Arg. This assumption was confirmed by the shift of the signal at m/z 1021, corresponding to the loss of Arg (Δm=156) from the C-terminus of the peptide, following carboxypeptidase B digestion [34]. The molecular mass was measured as 16099.90±1.69 and excluded the occurrence of any other replacement. In fact, the calculated molecular mass of the protein, taking into account the substitutions Arg for Lys at position 144 into the ovine β^B sequence, is 16101.46 daltons, in agreement with the experimental results.

The autoradiograms of mouflon and sheep DNAs digested with BamHI and probed with the goat ϵ^{IV} gene are shown in Fig. 2. It is clearly evident from the restriction patterns that mouflon with Hb B and domestic sheep with Hb A are characterized by fragments of 9.0, 5.5, and 6.6 kb containing the ε pair genes of the juvenile (ϵ^I and ϵ^{II}), of the adult (ϵ^{III} and ϵ^{IV}), and of the fetal set (ϵ^V ϵ^{VI}), respectively [32], so that they have to be considered as being homozygous for the triplicated haplotype. On the other hand, restriction patterns of domestic sheep with Hb B and Cyprus mouflon are characterized by fragments of 5.7 and 6.6 kb that were previously reported to contain the ϵ^I and ϵ^{II} pair of the adult set and the ϵ^{III} and ϵ^{IV} pair of the fetal set, respectively [1, 32]. Thus, Cyprus mouflon, as well as sheep of the Hb B type, is homozygous for the duplicated haplotype and carries a β-globin gene which should be considered as being of the "sheep β^B type". The organization of the β-globin gene clusters found in sheep and in mouflon as deduced by digestion of DNA by BamHI, EcoRI and HindIII restriction enzymes is presented schematically in Fig. 3, where locations of restriction sites are according to Garner and Lingrel [1, 2], Townes et al. [3], Di Gregorio et al. [11], and Rando et al. [32].

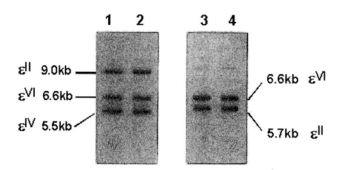

Fig. 2. Autoradiograms of mouflon and sheep DNAs digested with Bam HI enzyme and probed with goat ϵ^{IV} gene. The nomenclature of the gene is according to Garner and Lingrel [1, 2]

The high homology existing between β^Cyprus and sheep β^B genes was confirmed by sequencing of the two PCR-amplified genes, which showed only one nucleotide difference within the three coding regions. As shown in Fig. 4, this is at the level of exon 3, codon 144: A*A*A (coding for Lys) is present in sheep β^B-gene, whereas A*G*A (coding for Arg) is present in Cyprus mouflon gene. This is in complete agreement with sequence obtained by the FAB/MS methodology. The complete sequence of the β^Cyprus gene is shown in Fig. 5. Four additional differences out of 1603 nucleotides (0.25%) were observed within the noncoding regions.

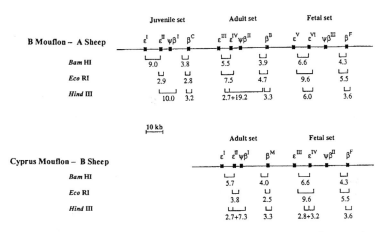

Fig. 3. The organization of the β-globin gene clusters of Cyprus mouflon, Sardinian mouflon of the Hb B type, and sheep of the Hb A and Hb B type. Restriction fragments were obtained by hybridizing the DNA with goat ε^IV and β^F probes. Mouflon of the Hb B type and sheep of the Hb A type show fragments of identical length on a triplicated *locus* chromosome. Sheep of the Hb B type and Cyprus mouflon show fragment of identical length on a duplicated *locus* chromosome. Lengths of fragments are given in kilobases

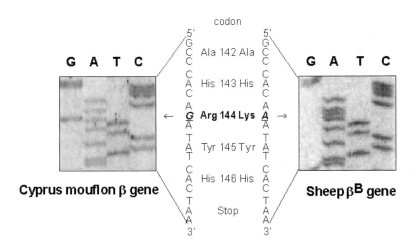

Fig. 4. Sequencing of the sheep β^B and Cyprus mouflon β-globin gene showing the A*G*A→A*A*A nucleotide change at codon 144 corresponding to the Arg→Lys amino acid substitution

```
                            20                  40                  60
SβB   aaaaatcctttccattttctgaagcccggattcttcatttatgtaataagaaaattgagg
βCyp  ............................................................

                            80                  100                 120
SβB   aggtaggtttccaagaggttacctcgttgtgactctaaatctctacaagcaaacttgct
βCyp  ............................................................

                            140                 160                 180
SβB   aaggaagatgattttagtagcaatgtgtattgctggaatgactgagaccttgagatgccc
βCyp  ............................................................

                            200                 220                 240
SβB   agaaagagggctgatggtctaaagtcagtgccaggaagaccaagtagaggtatggctatc
βCyp  ............................................................

                            260                 280                 300
SβB   accattcaagcctcaccctgtggaaccacaacttggcacgagccaatctgctcacagaag
βCyp  .........................................................

                            320                 340                 +1
SβB   cagggagggcaggaggcagggctgggcataaaaggaagagccgggccagctgccgcttac
βCyp  ............................................................
                                                                    420
                            380                 400      MetLeuThrAl
SβB   acttgcttctgacacaccgtgctcactagcagctgcacaaacacacaccATGCTGACTGC
βCyp  ..............................................................
                            440                 460                 480
      aGluGluLysAlaAlaValThrGlyPheTrpGlyLysValLysValAspGluValGlyAl
SβB   TGAGGAGAAGGCTGCCGTCACCGGCTTCTGGGGCAAGGTGAAAGTGGATGAAGTTGGTGC
βCyp  ............................................................

      aGluAlaLeuGlyAr     500                 520                 540
SβB   TGAGGCCCTGGGCAGgtaggtatcccacttacaagacaggtttaaggagagtgaatggca
βCyp  ............................................................
                            560                 580                 600
SβB   cctaggcatgcagggacagagctgtccctgagattctgaaagctgctgacttcctctgac
βCyp  ............................................................
                                                  640                 660
                            620 gLeuLeuValValTyrProTrpThrGlnArgPhePhe
SβB   cttgtgctgtttttctcccccttagGCTGCTGGTTGTCTACCCCTGGACTCAGAGGTTCTTT
βCyp  ..............................................................
                            680                 700                 720
      GluHisPheGlyAspLeuSerAsnAlaAspAlaValMetAsnAsnProLysValLysAla
SβB   GAGCACTTTGGGGACTTGTCCAATGCTGATGCTGTTATGAACAACCCTAAGGTGAAGGCC
βCyp  ............................................................
                            740                 760                 780
      HisGlyLysLysValLeuAspSerPheSerAsnGlyMetLysHisLeuAspAspLeuLys
SβB   CATGGCAAGAAGGTGCTAGACTCCTTTAGTAACGGCATGAAGCATCTCGATGACCTCAAG
βCyp  ............................................................
                            800                 820                 840
      GlyThrPheAlaGlnLeuSerGluLeuHisCysAspLysLeuHisValAspProGluAsn
SβB   GGCACCTTTGCTCAGCTGAGTGAGCTGCACTGTGATAAGCTGCACGTGGATCCTGAGAAC
βCyp  ............................................................

      PheArg              860                 880                 900
SβB   TTCAGGgtgagtttgtggagtcctcaatgttttccttcttcttttttatggtcaagctgat
βCyp  ...........................................................c..
```

Fig. 5. Nucleotide sequence of the sheep βB- (SβB) and Cyprus mouflon (βCyp) β-globin genes from position -358 to +1683 with respect to the Cap site. The coding sequences, in *capital letters*, are indicated with the three-letter amino acid code. The promoter sequences, Cap site, and polyadenylation signal are *underlined*. Deletions are marked with *hyphens*

```
                  920             940             960
SβB   gttatggggagaaggctgaatgacaggacacagtttagaatggagaagaggtattctggt
βCyp  ..............t.............................................
                  980             1000            1020
SβB   tagagtgctaaggactcctcagaaccgtttagactcttttaacctctgctcacaaccatc
βCyp  ............................................................
                  1040            1060            1080
SβB   atttcctctgattcattcttgttctctgttgtctgcaatgtcctcttttttagttatactt
βCyp  ............................................................
                  1100            1120            1140
SβB   tttattttgaggggtttaatttgaaaaaaaaaatttattttatcaactttaaaaatcatatc
βCyp  ............................................................
                  1160            1180            1200
SβB   taatattttccccttatctgtttctttcaaggaataaatgttctattgctttttgaaatg
βCyp  ............................................................
                  1220            1240            1260
SβB   attcaaaatgataaaaatgataacaagttctggattaaaaagagagaaacatttctaaac
βCyp  ............................................................
                  1280            1300            1320
SβB   atatattcaggaagacataggtagatacacatcagtagtaacatcttcgcttcagtcatc
βCyp  ............................................................
                  1340            1360            1380
SβB   cttgtgcttatattctacggtcacagcttgggatgagactgaaataccctgaatctaacc
βCyp  ........g....-..............................................
                  1400            1420            1440
SβB   ttggacttctctcatagctcagttggtaaagagtctgcctgcagtgcaggagatcccagt
βCyp  ............................................................
                  1460            1480            1500
SβB   tcgattcctgggtcaggaagaatggctggagaagggataggctacccactccagtattct
βCyp  ............................................................
                  1520            1540            1560
SβB   tgtgcttcccttgtggctcagctggtaaagaatctgcctgcagtgcgggagacctgggtt
βCyp  ............................................................
                  1580            1600            1620
SβB   cttctatccatgggttgggaagatcccctggagaagggaaaggctaccctctccagtatt
βCyp  ............................................................
                  1640            1660            1680
SβB   ctggcctggagaaatccgtggactgtatagtccatgggggttgcaaagagtcagacatgac
βCyp  ............................................................
                  1700            1720            1740
SβB   tgagcaactttcactttactaacctgcactaaccctgcccttgcttaatgtcttttccac
βCyp  ............................................................
                  1760            1780            1800
      LeuLeuGlyAsnValLeuValValValLeuAlaArgHisHisGlyAsnGluPheTh
SβB   acagCTCCTGGGCAACGTGCTGGTGGTTGTGCTGGCTCGCCACCATGGCAATGAATTCAC
βCyp  ............................................................
                  1820            1840            1860
      rProValLeuGlnAlaAspPheGlnLysValValAlaGlyValAlaAsnAlaLeuAlaHi
SβB   CCCGGTGCTGCAGGCTGACTTTCAGAAGGTGGTGGCTGGTGTTGCCAATGCCCTGGCCCA
βCyp  ............................................................

      sLysTyrHisStop  1880            1900            1920
SβB   CAAATATCACTAAgctccccttcctgatttccaggaaaggttttttcatcctcagagccc
βCyp  ..G.........................................................
      Arg
                  1940            1960            1980
SβB   aaaaattgaatatggaaaaattatgaagcattttgagcatctggcctctgcttaataaag
βCyp  ............................................................
                  2000            2020            2040
SβB   acactttttctcattgcactggtgtatttaaattatttcactgtctcttactcagatgggc
βCyp  ............................................................
```

Discussion

The data collected indicate that a high degree of homology exists between Cyprus mouflon β-globin gene and sheep β^B-globin gene, thus indicating the close genetic similarity between wild and domestic sheep. Though this homology, at the level of Hb and globin genes, has been already pointed out [22, 23, 26, 32, 35], it seems worth observing that the two mouflon varieties studied extensively so far (Corsico-Sardinian and Cyprus) exhibit two different chromosomal organizations of the β-globin gene cluster together with different structures of the β-globin chain. The Sardinian mouflon is almost monomorphic (*f*=0.94) for the so-called "β^B-chain", which is linked to the triplicated globin gene set arrangement corresponding to sheep of the Hb A type and goat [22, 23, 26, 32], whereas the mouflon living on the island of Cyprus is entirely monomorphic (*f*=1) for the duplicated gene set arrangement corresponding to sheep of the Hb B type. It is not surprising to observe that the homologous β^{Cyprus} and sheep β^B differ only in an amino acid at position 144, due to a single nucleotide change, and that the homologous mouflon β^B and sheep β^A differ in two positions (75 and 76). Among others, the arginine residue at position β 144, corresponding to HC1 terminal, is a constant residue in ruminants whereas lysine at the same position is constant in primates [36]. It may be, therefore, that the two amino acids are of importance in determining the oxygen affinity of the Hb tetramer. It has been reported that sheep Hb B (having Lys at 144) display a lower oxygen affinity than mouflon Hb B and sheep Hb A (both having Arg at that position) [25]. The low-oxygen level, mountainous habitat of the Cyprus mouflon requires a Hb with a relatively high oxygen affinity, which implies that its "sheep B" type Hb should have an oxygen affinity higher than sheep Hb B and mouflon Hb M, due to the presence of Arg at position 144. This might be considered an example of molecular adaptation, which might be essential for the well-being of this island species. Further studies on the oxygen-binding capacity of the mouflon Hb are planned in order to better understand the phylogenetics, evolution, and adaptation to the environment of ovines. The results of this study again support the notion that the nomenclature of ovi-caprine Hbs and globin genes is confusing and misleading. Even though the gene described here is a "sheep β^B-type" and not a "mouflon β^B-type", we prefer to call this new type of mouflon Hb "Hb Cyprus" and the β-globin gene "β^{Cyprus}" gene. A different nomenclature, such as the "β$^{B-Cyprus}$" or something similar, would cause additional confusion. However, the different chromosomal organization exhibited by the Corsico-Sardinian and the Cyprus mouflon and the molecular adaptation of the latter is an indication that the Corsico-Sardinian and the Cyprus mouflons are not absolutely identical as Pfeffer stated [15]. These differences, together with the differences in horn shape and size, body colors and body size, support the suggestion that the Cyprus mouflon should be kept as a distinct subspecies [20].

Acknowledgements

This work was supported in part by INTERREG II and Structural Funds of the European Community, and refers parts of the material described by E. Serreri in her thesis for Dottorato di Ricerca degree of the University of Sassari, Italy.

References

1. Garner KJ, Lingrel JB (1988) Structural organization of the β-globin locus of B-haplotype sheep. Mol Biol Evol 5:134-140
2. Garner KJ, Lingrel JB (1989) A comparison of the β^A- and β^B-globin gene clusters of sheep. J Mol Evol 28:175-184
3. Townes TM, Fitzgerald MC, Lingrel JB (1984) Triplication of a four-gene set during evolution of the goat β-globin locus produced three genes now expressed differentially during development. Proc Natl Acad Sci USA 81:6589-6593
4. Huisman THJ, Adams HR, Dimmock MO, Edwards WE, Wilson JB (1967) The structure of goat hemoglobins. J Biol Chem 242:2534-2541
5. Huisman THJ, Lewis JP, Blunt MH, Adams HR, Edwards WE, Wilson JB (1969) Hemoglobin C in newborn sheep and goats: a possible explanation for its function and biosynthesis. Pediatr Res 3:189-198
6. Boyer SH, Crosby EF, Noyes AN, Kaneko JJ, Keeton K, Zinkl J (1968) Hemoglobin switching in non-anemic sheep. Johns Hopkins Med J 123:92-94
7. Bunch TD, Foote WC, Spillett JJ (1976) Translocations of acrocentric chromosomes and their implications in the evolution of sheep (*Ovis*). Cytogenet Cell Genet 17:122-136
8. Bunch TD, Nadler, CF (1980) Giemsa-band patterns of the tahr and chromosomal evolution of the tribe Caprini. J Hered 71:110-116
9. Bunch TD, Nguyen TC (1982) Blood group comparisons between European mouflon sheep and North American desert bighorn sheep. J Hered 73:112-114
10. Ryder ML (1984) Sheep. In: Mason IL (ed) Evolution of domesticated animals. Longman, London, pp 63-85
11. Di Gregorio P, Rando A, Masina P (1987) Differences in the DNA restriction patterns between sheep with Hb A and Hb B. Anim Genet 18:214-247
12. Poplin F (1979) Origine du muflon de Corse dans une nouvelle perspective paléontologique, par marronage. Ann Genet Sel Anim 11:133-143
13. Vigne JD (1983) Les mammiferes post-glaciaires de Corse. Étude archéozoologique. Gallia Prehistoire Éditions du CNRS, Paris
14. Masseti MG (1997) The prehistorical diffusion of the Asiatic mouflon *Ovis gmelini* Blyth, 1841, and of the Bezoar goat, *Capra aegagrus* Erxleben, 1777, in the Mediterranean region beyond their natural distributions. In: Hadjisterkotis E (ed) The Mediterranean mouflon: management, genetics and conservation. Cassoulides and Sons, Nicosia, pp 1-19
15. Pfeffer P (1967) Le mouflon de Corse (*Ovis ammon musimon* Schreber, 1782). Position systematique, écologie, et éthologie comparées. Mammalia 31:1-262
16. Haaften JL Van (1971) Study on the situation of the mouflon in Cyprus and Turkey. CE/Nat 19:1-12
17. Schaller GB (1977) Mountain monarchs. Chicago University Press, Chicago and London.
18. Cugnasse JM (1994) Révision taxinomique des mouflons des îles méditerranéennes. Mammalia 58:507-512

19. Hadjisterkotis E, Bider JR (1997) Cyprus. In: Shackleton DM (ed) Wild sheep and goats and their relatives. Status survey and conservation action plan for Caprinae. IUCN, Gland and Cambridge, pp 89-92

20. Hadjisterkotis E, (1996) Herkunft, Taxonomie und neuere Entwicklung desw Zyprischen Muflons (*Ovis gmelini ophion*). Z Jagdwiss 42:104-110

21. Valdez R, Nadler CF, Bunch TD (1978) Evolution of wild sheep in Iran. Evolution 32:56-72

22. Naitana S, Ledda S, Cocco E, Manca L, Masala B (1990) Hemoglobin phenotype of the European mouflon sheep living on the Island of Sardinia. Anim Genet 21:67-75

23. Rando A, Di Gregorio P, Capuano M, Senese C, Manca L, Naitana S, Masala B (1996) A comparison between the β-globin gene clusters of sheep (*Ovis aries*) and Sardinian mouflon (*Ovis gmelini musimon*). Genet Select Evol 28:217-222

24. Blunt MH, Huisman THJ (1975) The hemoglobin of sheep. In: Blunt MH (ed) The blood of sheep. Springer, Berlin Heidelberg New York, pp 155-183

25. Corda M, Giardina B, Pellegrini M, Manca L, Olianas A, Sanna MT, Fais A, Masala B (1997) A comparative study on the functional properties of the wild European mouflon and domestic sheep hemoglobins. Comp Biochem Physiol 117B:417-420

26. Rando A, Di Gregorio P, Capuano M, Senese C, Hadjisterkotis H, Musino L, Palici di Suni M, Manca L, Masala B (1997) The β-globin gene clusters of domestic sheep (*Ovis aries*), and of Sardinian (*O. gmelini musimon*) and Cyprus (*O. g. ophion*) mouflon. In: Hadjisterkotis E (ed) The Mediterranean mouflon: management, genetics and conservation. Cassoulides and Son, Nicosia, pp 67-72

27. Masala B, Manca L (1991) Detection of the common Hb F Sardinia [$^A\gamma^{75(E19)Ile \to Thr}$] variant by isoelectric focusing in normal newborn and in adult affected by elevated fetal hemoglobin syndromes. Clin Chim Acta 198:195-202

28. Manca L, Formato M, Demuro P, Pilo G, Gallisai D, Orzalesi M, Masala B (1986) The γ globin chain heterogeneity of the Sardinian newborn baby. Hemoglobin 10:519-528

29. Masala B, Manca L (1994) Separation of globin chains by the reversed-phase high-performance liquid chromatography. Methods Enzymol 231:21-44

30. Pucci P, Carestia C, Fioretti G, Mastrobuoni AM, Pagano L (1985) Protein fingerprint by fast atom bombardment mass spectrometry: characterization of normal and variant human haemoglobins. Biochem Biophys Res Commun 130:84-90

31. Sambrook J, Fritsch EF, Maniatis T (1989) Molecular cloning. A laboratory manual. Cold Spring Harbor Laboratory, New York

32. Rando A, Di Gregorio P, Masina P (1989) Differences in the number of embryonic and pseudo-beta-globin genes between Hb A and Hb B sheep. Biochem Genet 27:91-98

33. Sanger F, Nickeln S, Coulson (1977) DNA sequencing with chain terminating inhibitors. Proc Natl Acad Sci USA 74:5463-5467

34. Pucci P, Malori A, Marino G, Metafora S, Esposito C, Porta R (1988) Beta-endorphin modification by transglutaminase in vitro: identification by FAB/MS of glutamine-11 and lysine-29 as acyl donor and acceptor sites. Biochem Biophys Res Commun 154:735-740

35. Stratil A, Bobák P (1988) Comparison of biochemical polymorphisms in mouflon and sheep: Isoelectric differences in haemoglobins and quantitative variation of mouflon haemopexin. Comp Biochem Physiol 90B:159-162

36. Poyart C, Wajcman H, Kister J (1992) Molecular adaptation of hemoglobin function in mammals. Respir Physiol 90:3-17

Subject Index

DATE DUE

MAY 2 5 2002	
APR 2 8 2003	
JUN 2 7 2005	
MAY 1 6 2007	